Perspectives on Technology and Culture

Perspectives on Technology and Culture

by Egbert Schuurman

translated by John H. Kok

Dordt College Press
Sioux Center, Iowa

This book was translated from the Dutch by John H. Kok.

Printed in the United States of America

Published simultaneously in South Africa by the
Institute for Reformational Study,
Potchefstroom University for Christian Higher Education.

Dordt College Press
498 Fourth Avenue NE
Sioux Center, Iowa 51250

ISBN: 0-932914-33-0

Preface

The philosophy of technology is a new shoot on the tree of philosophy. Modern technology is relatively young. Classical technology is much older. We will see that technology changes when it is given a scientific basis and the scientific method is introduced. When technology is grounded in science we speak of modern technology. A philosophy of technology focuses mainly on the connection between science and technology.

Chapter I addresses questions relating to the background of modern technology, showing the influence of science in the phenomenal development of technology. Many cultural philosophers focus on the fact that modern technology's potential and threat lie in the central role that science plays in it today.

Chapter II discusses philosophical questions pertaining to the technological sciences and their methods, analyzing the influence of science on technology and the structure of the method of scientific design. Attention is also given to the recent rise of biotechnology.

Chapter III addresses philosophical and ethical problems and issues relating to the technological sciences and, in that connection, to modern technology. The central theme of these final paragraphs is that science ought to stand in a relationship of service to meaningful technology.

Chapter IV deals with the ethical and philosophical implications of current issues like information technology, biotechnology, and genetic manipulation. The final chapter underscores the need for a Christian approach when assessing contemporary culture. A selective bibliography used in writing this study is included at the end.

I would like to thank my translator, John Kok, for his fine job. He and I would like to thank Charles Adams and Mike Vanden Bosch for reading the translation in manuscript form and making helpful suggestions.

I hope this study promotes critical and creative reflection on modern technology, its history, structure and problems.

Egbert Schuurman
August, 1994
The Netherlands

Table of Contents

Chapter III. Philosophical-Ethical Questions

Chapter IV. Current Issues

Chapter V. Technological Culture between the Times

Chapter One

BASIC QUESTIONS AND BACKGROUND

1. Technology and the Technological Sciences

Technology is as old as humankind. Originally it included only human handicraft and was part of the work done in the home. As such it was sensitive to immediate and local needs. One worked with modest tools and instruments that were seldom more complicated than a simple loom, millstone, or lathe. The energy required for this labor was supplied by humans or animals. Technological inventions like the wheel, the lathe, the potter's wheel, the water wheel, the smelting furnace, etc. were developed only to the extent that human or animal propulsion allowed.

Technological development did surge here and there, for example, during the rule of Rome, in the execution of public works, or in the waging of war. Only when science began to work closely with technology and the industrial enterprise was there an explosive development of technology.

This book studies the relationship between science and technology. The scientific foundation of technology has many consequences. Science solves technological problems systematically. It analyzes nature with an eye to its possible technological functions. Scientific methodology can solve many problems quickly and efficiently. These solutions are recorded, remain scientifically adaptable, and can therefore be applied in new and better ways. Electronics, chemical technology, material science, aerospace engineering and technology, maritime technology, mines engineering, or civil engineering and technology, are unimaginable without the influence of science.

Years ago design and implementation were undivided. Due to the influence of science on technology these have become two distinct phases. Especially in the preparatory phases of design, for example, the influence of science is unmistakable.

Literally, "technology" or "technological science" involves the scientific knowledge of *technique*. Technological science and technology are closely interwoven with each other. Given the wide variety of sectors within technological science we will also speak of "the technological sciences," each of which corresponds to a certain field of technology. Paragraph 18 is devoted to an overview of the "technological sciences."

2. Philosophy of Technology and the Technological Sciences

Technological science is not the same as technology. Both are, however, inseparably connected to each other. Technological science focuses on how something ought to be made. Technology and the technicians make it. The technical scientist is usually also a technician who applies the science or who further develops the science on the basis of the application. But a sharp distinction between technological science and technology is not possible. New developments in technological science go hand in hand with new technological possibilities. New inventions are significant for technological science and, as we will see later, also for the natural sciences.

Philosophical reflection on technological science is, as a result, also dependent on a philosophical consideration of technology. That is to say: a philosophy of technological science is inconceivable without a philosophy of technology. This book strongly emphasizes the mutual relationship of these two. And yet, distinctions can be made. The remainder of this chapter is primarily devoted to a philosophical consideration of technology. We have to know precisely what "technology" is, how in overview technology has developed, which forces have driven it, the rich diversity of tools and methods that characterize it, what its possibilities and limits are, etc. But because technology, due to the influence of science, has become a pre-eminent cultural power to which we are deeply indebted, philosophical consideration requires that we examine the inner connection of science and technology. A survey of the various currents in that philosophical reflection (in §8 and following) brings home the fact that the possibilities and dangers of modern technology have everything to do with the relationship of science and technology. This survey forms the transition to the philosophy of the technological sciences, which is the topic of Chapter 2.

Because the dangers and threats of modern technology can be traced to an irresponsible connection between science and technology, it is unavoidable that the ethical dimension will surface in the elaboration of various philosophical currents. After a philosophical discussion of the technological sciences in Chapter 2, we will return to the matter of ethics in Chapter 4. There we will want to clarify how science and technology can best be connected with each other so that the problems, threats, and dangers can, if not be avoided completely, at least be held within certain bounds.

3. What is Technology?

Philosophical reflection on technology and technological science must first answer the question, "What is meant by *technology*?" The words "technique" (or "technic") and "technology" are derived from the Greek word *technè*, which has an array of meanings. It can express all human (cultural) activity in distinction from nature as *physis*. But it signifies especially "skill" and "proficiency." It can also connote "cunning." *Technè* also proves to be related to the words *epistèmè*, meaning "understanding" and "(artistic) skill," and *poièsis*, that denotes "doing," "working," "creating," and also once again "skill." Obviously, the Greeks did not clearly differentiate between technology, science, and art. In other words, the etymology of the word "technique" is not going to be of much help in answering the question as to what technology is.

In their common usage "technique" and "technology" have, as well, a variety of meanings. We use the word technique when we want to talk about a certain kind of human activity and its result, but also generally when we have our eye on human proficiency and the methods it involves. Piano technique is an example of the first use; social techniques an example of the second. These ambiguities can hardly provide a basis for theoretical analysis. Any philosophical reflection on technology and technological sciences will have to ascertain with some precision what technology is.

Sound insight as to what we are discussing will prevent misconceptions and erroneous views of development. Unfortunately, philosophers themselves use these terms in a variety of ways. The French philosopher Jacques Ellul speaks about technique as a scientific method of domination. José Ortega y Gasset, the Spanish philosopher of culture, used the terms "technology" and "science" interchangeably. For Martin Heidegger, "technology" includes many cultural expressions, like machines, bureaucracy, information, and organization. The distinction between science and technology is not at all clear, given these uses. And yet the distinction does refer to a decided difference. The aim of science is always to gain knowledge of reality, while the goal of technology is to change reality. In like manner, we speak of "discovery" in science and of "invention" in technology. Science is the power of knowledge; technology is the power of formation. In other words, in science it is a matter of *knowing*, in technology a matter of *making* or *forming*.

In this study we take *technology* to be *the human formation of nature with the help of tools for human purposes*. We have to remember, however, that the nature of tools has changed substantially in the course of time due to the influence of science (see §5). This is so evident that technology is often simply referred to as *an applied natural science*. While this rightly emphasizes the significance of the natural sciences for technology (see §10), modern technology is itself primarily not a matter of knowing, but, as we said, of forming. Besides that, instead of saying that technology is applied natural science, it could better be said that modern technology is *applied technological science*. We will refer to the scientific, theoretical knowledge about the technological activity of formation as "the science of technology." This science can rightly be called a practical science.

"The science of technology" is usually equated with "technology." We will want to distinguish these two so as to avoid confusion. Technology and the science that deals with technology are not one and the same. That people confuse the two is more than accidental. Science has often been taken to be a technological instrument of domination. As we will see (§§21-2), it is precisely this instrumentalistic view of science, and of technological science in particular, that has given rise to ethical problems.

The power of human formation through the use of tools, with which human beings invent, design, construct, and fabricate new things and processes as the fruit of productive imagination, has changed substantially due to the influence of science. That is why, to distinguish it from the classical technologies of handicraft and guild, we speak of *modern* (science based) *technology*.

Before turning to the differences between guild or handicraft technology and modern technology, we must mention the fact that modern technology, in the sense of the power of formation, has in the recent past developed beyond tools. Biotechnology is one such example, in which scientists control life-processes or manipulate the genetic structure of micro-organisms, plants, animals, and even humans. In order to distinguish this kind of technology from *tool technology*, we will also want to talk about *biotechnology* and the science of the same (see §19).

Another misunderstanding deserves attention. When we say that technology is the power of formation through the use of tools, we are clearly distinguishing the tool from what is being made. The product of technology can itself be of a technological nature. For example, technology can produce new technological tools. But usually its products have another purpose. An automobile, a house, a church, a rocket: none of these have a technological purpose. All of these are results of technology, but they do not have as their purpose to further the development of technology, at least not directly.

4. History of Technology

To get a correct sense of its development, to gain insight into the influence of science on technology, and especially to be able to direct technological development responsibly, we must know the history of technology, and particularly of the spiritual forces that drive technology.

The history of technology can be subdivided into three main periods. During the first period technology developed very slowly. This phase, the *traditional era*, lasted until about 1500 A.D. We refer to the period from 1500 to 1800 as the practical era. Given the predominant influence of science on technology, we can call the next age the *era of modern technology*.

During the traditional era technological development remained static. This can be attributed to a lack of technological possibilities, but even more to deeper lying reasons. In the traditional era human beings were not motivated to change the practice, activity, and constructions of everyday life, but to *contemplate* nobler things. Inventions were made on paper, by Archimedes for example, but because of a dearth of technological motivation they were seldom implemented. For religious reasons, among which a devaluation of earthly life was predominant, work and technology were not encouraged. That is why the technology of the traditional era is more or less limited to things like the techniques of war, technological products connected with worship, and technology as a kind of intellectual free play.

The rise of the practical era is closely related to the spread of Christianity and the motivation to form culture that accompanies it. People are called to serve God in his creation and to help unfold it. The cultural mandate of Genesis impels humans to develop and value technology positively. This Christian motivation did not immediately develop a culture or dynamically unfold history. That came only after 1500. Some of the obstacles were spiritual. It took time to see what Christendom implied for culture, how it should influence practice. At first this process moved very slowly, and often very superficially, due to the wholesale conversion of entire peoples. Christianity also was often accepted piecemeal. The characteristic influence of the biblical message was often restricted to "supernatural priorities" and the realm of the institutional church. In addition, socio-cultural developments, like the fall of the Roman empire in the fourth century, hindered a prompt flourishing of technology. Later the rise of Islam deterred sea trade and cultural development stagnated.

Social changes and new spiritual currents between 1200 and 1600 prompted the rise of the New Age, the onset of a dynamic period for culture and history, and of the practical era for technology. To understand this development, we should best not think of "cause and effect," but of a multifaceted process of interaction. However, the spiritual motivation, understood in a religious sense, was the fundamental reason for the genesis of the new development.

Circumstances in society, like the upswing in trade due to the crusades, the developing craftsmanship in the guilds, and the social emancipation of the common folk, stimulated a developing technology. The growing strength of this liberation movement contributed to the disbanding of the guilds, which in turn contributed to renewing technology.

The history of technology is more complex than is often thought. Christendom itself was of considerable significance. A spiritual renewal and at the same time the secularizing of that same Christendom also supplied a powerful impetus.

From 1300 to 1500 mounting spiritual movements imparted a new attitude that would eventually split in two: on the one hand, the movement of the Renaissance and, on the other, that of the Reformation. Both of them address the place of man in the world. The Reformation focused on the human being as called by God to serve him in all things. For the Renaissance the theme of the self-empowerment of man began to predominate more and more. This motif of the Renaissance, amplified by the movement of humanism, extensively influenced philosophy and science. In the long run, after an initial flowering, the Reformation's influence on philosophy and science, in spite of its historically significant positive view of these pursuits, remained limited. The Reformation set its stamp on commerce, agriculture, and the development of trades. The spiritual movement of the Renaissance, intensified by the Enlightenment, became the moving force in the development of science and technology.

The practical era was ushered in by new developments in the technology used by craftsmen. Many inventions came into play, often one right after another. Even academics, like Leonardo Da Vinci and Simon Stevin, joined the ranks of inventive craftsmen.

Inventions like gun powder, typography, compass, the mechanical timepiece, and many of the instruments needed for scientific research helped turn the page around 1800 from the practical technology of craftsmen to the era of *modern* scientifically based *technology* (Forbes, 1963).

From the eighteenth century on, the spirit of the Enlightenment shaped science and modern technology. In organizing society it let science lead. As a result, and with the rise of modern technology, culture became secularized.

An interesting detail deserves mention. More than one scholar has pointed out that the later motif of the Enlightenment had begun to develop already at the end of the Middle Ages. The Enlightenment made the claim that mankind, independently empowered by the light of reason and with the help of science, would be able to control the future. Some said the influence of magic on developing technology was the precursor of this Enlightenment pretension. People turned to alchemy as life's cure-all, hoping that it might even be able to keep mortals from death. The same spirit is evident in its search for that primal material from which gold could be fashioned and one's financial independence guaranteed. Attempts to transcend the bounds of space and time and to conquer the law of cause and effect likewise manifest a motif that, modified by the admixture of science, became even more preponderant in the Enlightenment.

In summary, the spiritual forces shaping the development of modern technology are derived from the Reformation and from the Renaissance, later transposed into the Enlightenment. Both movements confirm the influence of science on technology. However, the direction of each is different. Only now that we are confronted with a diversity of problems, tensions, and threats within and around modern technology are we more and more becoming aware of this divergence in direction. That is also the reason why unanimity is lacking today as to the road to follow in the future. As a result we can speak of a crisis in the development of technology. For that reason, too, ethical reflection is inescapable.

5. Technology Then and Now

The history of technology teaches very clearly that technology has to do primarily with *tools* and their use. The history of tools discloses the gradual growth of human formative mastery; a growth that today has resulted in the imposing power of puny people—puny compared to the means at their disposal.

This development is strikingly evident when we compare the classical technology of the tradesman and craftsman with modern technology. In making the *comparison* we can leave the actual development out of the picture for the moment. We talked about that in the previous paragraph. We will limit ourselves, point for point, to differences in *milieu*, materials, energy, skill, the sequence of the technological operations, cooperation within technology, work methods, mass production, and the nature of technological development.

(a) In the era of classical technology man was by and large surrounded by a natural milieu. He was basically limited to what was allowed and supplied by nature. There are only incidental examples of an ability to distance oneself from nature. In contrast, people in the modern era live in a milieu already formed by technology. Highly developed tools are readily available and the distance from nature is on the increase.

(b) It used to be that people had access only to *materials* that were readily available in nature. In the modern era the desire is to depend on natural materials as little as possible. Natural materials of the same kind still differ too much in particulars and are usually mixed with natural "impurities." In modern technology materials are expected to be fully refined and have a uniform consistency. An extreme example of this development, in which the technician dictates the desired characteristics, is commonplace in materials technology. One can think, for example, of very light materials that can at the same time withstand very intense forces, of nonabrasive materials subject to little or no wear, and of modern synthetic materials, like plastics, that display a whole new set of characteristics.

(c) In the previous era the *energy* required for technological formation, for construction or fabrication, depended on the push and pull of animal and human muscle power. In modern technology energy is drawn directly from nature, from wind or water, or indirectly through the burning of coal and oil or even by splitting atoms. The direct use of the sun's energy via solar cells and panels is presently a reality.

(d) For centuries human *skills* in the use of simple tools defined the extent of *technological formation.* In modern technology these have been replaced by machines. Machines not only refine and enhance these skills. They also give rise to new possibilities. Electronics and computer technology are particularly obvious examples.

(e) As far as *technological operations* are concerned, craftsmen and tradesmen had the lead from beginning to end. In modern technology this process has been taken over by automation. The modern tool is an independently working, technological operator.

(f) In technology of the practical era there was a direct relationship between *producer* and *consumer*. Modern technology is characterized by a strongly differentiated *cooperation* between engineer, builder, and employee. In addition, production is usually interwoven with an industrial enterprise. There is no longer direct contact between producer and consumer.

(g) Pre-modern technology was, as it were, enfolded within nature. Modern technology is characterized by a *distance* with respect to nature because of the *scientific approach* of modern technology. By distancing itself from everyday life, the plan and its execution have become divorced from each other. The workers' place has become a scientifically calculated station. When technological formation and its execution become self-supporting in the process of automation, the worker is left without work. At the same time, with the scientific approach, the process of design is brought to a higher level, creating the possibility for new constructions and new inventions.

(h) When technological formation is automated, we have *mass production*. Products of the craftsman, in contrast, are unique and of a singular character.

(i) Earlier, technology was limited to the natural possibilities of man, in the sense of the creationally given measure of bodily exertion and capacity. It remained within parameters defined by the human hand and sense organs and was based on everyday knowledge. This is what makes the technology of the craftsman "natural." In this kind of technology there is little or no talk of development. The experienced rhythm of the routine was passed on from father to son, from master to apprentice. *Pre-modern technology is undifferentiated and static in character.*

Due to its scientific basis, *modern technology* has experienced a massive proliferation. It is *highly differentiated and dynamic in character.* In addition, the breadth of modern culture bears the stamp of technological development. A technological culture is usually vivacious. This cultural dynamism especially typifies the recent past. If we could condense the entire length of technological development into a period of twenty-four hours, then the invention of the steam engine would have taken place about thirty seconds ago. The most recent seconds would have contained tumultuous developments that unimaginably change the very countenance of the earth with lightening speed. In a material sense, life is enriched enormously. At the same time it becomes increasingly apparent that these changes are accompanied by colossal disadvantages. The extent to which science is to blame for this remains to be seen (see §21).

6. From Tool to Robot

To get a handle on the change that has taken place in *tools* over the course of time, we can best focus on the changing relationship between humans and their tools. What becomes apparent is that human beings gradually recede into the background and that tools more and more take over what used to be human tasks. Originally humans made or formed everything using tools. Later they directed and controlled the tools that did the making. Finally, today, tools have become technological operators, working on their own, with humans determining the process at a distance through design and programming.

This changing relation between human beings and their tools has been an on-going topic of discussion among philosophers of technology. Ernst Kapp, said to be the first philosopher of technology, introduced the subject already in 1877. He considered the development of technology to be an unconscious extension of human organs. Though unaware of the fact, we tend to make tools that amplify natural skills and abilities like seeing, hearing, grasping, and lifting.

Human beings with tools are in a position to do more than they are without them. The power and achievement of human labor increases as these tools become more advanced. Although there are exceptions, we can in a sense agree with Kapp that people have objectified or projected certain human functions in their tools. In doing so, the burden on human beings decreases while the human function in effect increases. Because of the influence of science, particularly of the natural sciences, information science, and the technological sciences, tool development is still experiencing a stage of rapid growth. An increase in tools that do their work with little or no human intervention has been the result.

Originally human beings supplied the energy required for using tools. Their skill, or lack of it, in using these tools determined the outcome. When tools become *instruments*, human skill is defined by predetermined limits. The instrument's construction allows human energy to achieve more than ever before. At the same time, the actual human energy exerted decreases to the extent that energy can be drawn from non-human sources: horse power, then wind and water power, and after that a long gradation of energy sources that reaches its limit with atomic energy. Though exerting less, humans today can do more. This is especially the case where, with the help of the natural sciences, energy is produced by technological formation.

During the next stage human beings still often have to service, steer, and supervise their tools. But in the process of turning to other sources of energy, the skill and proficiency are increasingly transferred to the *machine*, with generally positive results. Ultimately, using the principles of cybernetics, the regulation and supervision of the production process is also transferred to machines.

These last developments move us beyond the phases of *material technology* and *energy technology* and into the period of *information technology*. With the appearance of *computers* and *automatons* (e.g. robots) we stand on the threshold of a new era.

The age of information technology is characterized especially by the introduction of the *computer*. It is a fascinating technological achievement which, at the same time, breaks new ground. In the last thirty years, developments in computer technology have been almost overwhelming. We have moved from mammoth, unreliable, expensive, main-frame computers to small, dependable, relatively inexpensive, user-friendly machines that can process mounds of information very quickly.

With respect to technology, the rise of the computer generated a giant step toward complete automation. In cases of total automation the computer is completely integrated in the production process. Acronyms abound: CAD computer aided design; CAM computer aided manufacturing; CIM computer integrated manufacturing. Their presence pervades, for example, the tool and die industry and modern chemical processing companies. We see analogous developments in, for example, aerospace, communication and medical technology, in telecommunications, and in automation in the service sector. Developments in *micro-electronics*, including micro-computers and micro-chips, have accelerated this process. However, as a result of automation, increased unemployment is no longer limited to technicians, but now includes all fields of cultural labor that are systemized.

In all of these cases the computer is present as a *steering or control mechanism.* In the case of science the computer is used as a *tool for thought.*

More than any other tool, the computer has shown that science and technology reciprocally enrich each other. The computer as technological result serves science, while science is a perpetual impetus for technological development. The computer is not only important for technological science. Because of the applicability of logic and mathematics as scientific aids for (quantitative) analysis, the general use of the computer has become commonplace in scientific research. In addition, computers are used in fields like operations research, decision-making, and praxeology—the study of human action and conduct.

Emphasizing that the computer, too, is a tool can keep one from falling prey to all kinds of speculation with respect to "thinking machines," as though a computer could ever exist or work independent of human beings. To believe that, is to become the dupe of the computer. Conjectures about machines that can think too often forget that when it comes right down to it the computer is and will remain a tool of human beings. Computers can process very large quantities of information quickly and without error. They perform mathematical operations that are so involved that humans could not even begin to duplicate their calculations. And they generate answers that are more inclusive than humans, given their limited ability to survey all the available information, could ever produce without the computer. Nevertheless, one ought not to forget that the information stored in its memory and the programs it uses are where they are because humans put them there. In this sense the computer remains subservient to human beings, who in turn are responsible for the outcome. Even though computer results may be startling and their makers or users incapable of ever achieving as well, if at all, similar results without a computer, that does not mean that computers are not dependent on human beings. Computers are there to serve human beings. They help to analyze, search, recognize, calculate, sort, and, when it comes to technology, to control (Schuurman, 1982).

But now back to the continuing development of technology. In the development of industrial as well as nonindustrial *robots* the attempt is being made, with increasing success, to transfer the remaining functions of human beings to the technological operator. We can speak of a robot when the internal heart of the technological operator is a computer, the information available comes from external sensors as well as from the computer's program, and the external effects of the information processed run via artificial limbs. The robot should also be able to learn from its own "experience" and in so doing be able to better itself. It is understandable that where this combination of "sensitive" and "mental" functions presents itself people speak of a *robot with artificial intelligence*. However, some who use the term "artificial intelligence" are referring to *expert systems*.

In noting the place these systems occupy in the production process, we can correctly say that the history of technology shows that human tools have gone from things we can easily get a handle on and oversee to a *technological system* that is beyond our circumspection (Ellul, 1980).

Having come to the present in this long historical process, the dynamic development that characterizes modern technology is clear. And as though that were not enough, it still has many surprises in store for us. Robots are in a position to do even more than most can even imagine.

As a result, both optimists and pessimists speculate as to how technology will yet develop. On the one hand, the use of robots provokes very intense premonitions concerning the future: they will be used in ways presently unheard-of; wonderful promises are on their way to becoming reality. On the other hand, unanticipated results incite fear and anxiety. Will we be able to continue to oversee and control these new developments? Is the product going to consume the producer? Is science being misapplied or does the blame lie with science itself? How could that have happened?

7. Problems in the "Technological Culture"

Modern technology, whose activities and results are rooted in science, commences with James Watt in the second half of the eighteenth century. The industrial revolution follows in the nineteenth century, and in the twentieth century, at least according to Norbert Wiener, there emerges, riding on the shoulders of information technology, a second technological revolution.

Machine construction, chemical technology, electronics, transportation technology (automobiles and airplanes), materials technology, atomic energy, computer technology, aerospace technology, telecommunication, and biotechnology are developed one after the other and influenced and stimulated the one by the other. Education in the technological sciences and related disciplines skyrocketed as well. Some of the industrial arts that had been taught in high school were now given academic status at the college level, while university level institutes for technology were being founded as early as 1860.

Given the fact that we find ourselves right in the middle of the development of modern technology, we know not only of the advantages it has brought to man and society, but also of the disadvantages that accompany it. Modern technology, it would seem, has a questionable character.

For a long time the advantages of modern technology contributed to an uncritical attitude with respect to it. People spoke about the "triumph of technology," the "century of technology," even about the "miracles of technology." Many expected that the combination of science and technology would be the beginning of the end to humanity's problems. They had their eye on an increased life expectancy, overcoming most illness and transmissible defects, improving memory and intelligence through symbiosis with the computer, the automation of most activities, growth in the material standard of living, and the artificial generation of more suitable plants, organisms, and animals.

Needless to say, many of these are now reality. There is an uncanny dynamic to recent human history. Human beings are more powerful today than ever before. Technology is shoring up a level of material prosperity that is unparalleled in history. Moreover, leisure time is on the increase, such that many can liberate themselves and, for example, further themselves in many more ways than ever before.

Modern technology has put its stamp on culture. At the same time it functions as a basis for a phenomenal integration of humankind. An integral technological world culture is upon us.

However, it is time to attend as well to the problems, tensions, and threats that characterize the other side of the coin. People perceive these burdens to be the result of the growing technological power that has taken hold of the history of humankind at its core, imparting to it a fragility that forces us to ask the question whether present developments find us on the right road. In connection with this conundrum, there is even talk of a *crisis*. That is to say, there is no unanimity as to the solutions of these problems, which in the meantime are increasing.

I will focus briefly on a number of these problems. They have to do with (a) the relation between people and technology, (b) the relationship of people, technology, and nature, (c) the socio-economic context in which technology is developed, (d) questions concerning human responsibility, and (e) the relationship of technology and religion. We will see how much these problems coincide with *modern* technology and hence are connected to the influence exerted by science on technology. That will become particularly clear when, after reviewing these problems, we discuss different philosophical conceptions of the "technological culture."

a. The problem area that looms foremost concerns *the position of human beings in modern technology* as compared to their place previously. In the technology of the craftsman, human achievement is determined by the eye and hand, by a feel for the material, by imagination and a formative knack. Physical and mental abilities are brought to bear directly. People's work and their products are marked by quality, as well as by being distinct, one of a kind, and nonrepetitious.

In modern technology the situation is completely different. It is no longer a matter of a simple, personal, comprehensible technology, with few, if any, unforeseen consequences. Modern technology, in contrast, is very involved and characterized by an unoverseeable interaction, with, in many cases, unanticipated, radical repercussions. Within the arena of modern technology almost everything is quantifiable and characterized by an increase in tempo and range. Human labor has become a quantified business whose intensity is being stepped up continually. At the same time, the laborer is being taken up into an objective, impersonal, and dispassionate process. What was previously personal becomes matter of fact and vacuous. That helps to explain the callous attitude and indifference of many with respect to their work. People feel as if they are being sucked into an insurmountable, massive, mechanistic process; a process in which their energy is being exhausted. When this process becomes fully automated they are simply eliminated and can only look on, unemployed.

The designer of this technology, the *engineer*, experiences something very similar. In designing he has, with the computer, the most modern technological options at his disposal, and yet he often feels strapped. As a specialist he is increasingly dependent on other specialists. Given this dependence on others as well as on the computer, the individual engineer lacks the insight and capability to help direct the development of the technology.

For the laborer and engineer, as well as for the "outsider," the cutting edge of modern technology arouses apprehension. For example, the development of nuclear reactors promises a secure supply that can meet the increasing demand for energy, but we are at the same time oppressed by the many as yet unresolved technological and political risks that accompany these facilities, which can hardly be surveyed in their entirety. In addition to difficulties in controlling the technology—think of Three Mile Island and Chernobyl—problems concerning radioactive wastes continue to vex the minds of many. It remains to be seen whether the elementary forces of matter that are being released can be controlled sufficiently.

Something similar holds true for computer technology. It does not take too much to see that the computer works quickly and accurately and that what it generates will never fall outside of the programmed parameters. And yet we fear becoming increasingly dependent on the computer. Its results—albeit within certain limits (Schuurman, 1980)—have something dazzling about them. This is enhanced because the user and the programmer need not be one and the same person. As the users change, they are no longer cognizant of the criteria according to which the computer works, with the result that one can only trust that what it does it does well. This problematic situation is going to become more pressing as the capacity of computers increases and as machines are introduced that can learn or reproduce themselves. When one banks on so-called expert-systems and on robots with "artificial intelligence" (see §6), stagnation can arise because these systems can never completely adapt themselves to the dynamic change of reality (see §21).

A *computerocracy* seems to be just around the corner when these machines not only find their place within the production process, but also settle into the areas of economics, politics, and the like. Especially the manipulating of personal information stored in data banks, with its consequent manipulating of people, can be dangerous.

The perils of modern technology are manifestly evident when we note the dangers that are connected with the newest techniques of waging war, the pit of which is nuclear suicide.

The development of recombinant DNA technology—also called "genetic engineering"—is just one facet of biotechnology that next to its many surprises has as many imminent dangers in the offing (see §21).

b. A second problem area has to do with *how human beings relate to nature*. More than one scholar has pointed out that the earlier organistic view of nature has been replaced by a mechanistic conception. Nature is seen in terms of the physical categories of number, space, and causality. Measuring, weighing, and counting determines what is nature. The remainder is neglected. This mechanistic view of nature invites pervasive technological control in service of goals and desires set by men. Nature no longer seems to have its own value or meaning.

The means by which and the manner in which nature is controlled are given little or no attention in this utilitarian framework (see §21).

Early on, small-scale technology was still characterized by an integration with nature. That is no longer the case for modern technology. A rift has arisen between man and nature. Aided by modern technology human beings are intruding on nature on a large scale and with increasing rapidity. In doing so they undo the manifest coherence in nature. What is more, nature is not even given the opportunity to recover from the damage inflicted. All the while, unfavorable side effects of modern industrial technology, modern transportation, and what all too often approaches factory farming, continue to reinforce each other. The imminent collapse of the biosphere is one of the dangers. We use more from nature than it can produce and we leave more behind than it can decompose—not to mention what cannot be disintegrated. Because of the disruption that precipitates, nature when controlled technologically itself becomes a threat to humankind. That is evident from the greenhouse effect, changes in climate, a rising sea level, ozone depletion, acid rain, and desertification. All of these are the result of irresponsible procedures that have risked the coherence in nature and whose consequences can no longer be completely undone.

c. Technology is interwoven with *economic and political forces*. In addition to the difficulty of surveying the structure of modern technology, this intertwinement explains why the development of technology is experienced as a self-governing, *autonomous process* (Winner, 1977). The research and development of new technologies has often taken place (secretly) in multinational conglomerates and within the so-called military industrial complex. When government begins to play an ever increasing part in this process, experience shows that technological development embarks on an upward spiral. This growing *technocracy* is coupled with specialization and mass production. Technology responds, in the mean time, not only more to present needs and desires, but also artificially creates new ones. An inability to survey the whole, on the one hand, and redundancy, on the other hand, make it such that technological means not only have a devastating influence on nature, but at the same time constitute a direct threat to humanity. Because people more and more are becoming components of the technological system, philosophers like Karl Jaspers and Jacques Ellul have emphasized the inhuman character of modern technology. People are uprooted from their tradition and alienated from their culture. The gravity of the situation is often concealed by the numbing effects that emanate from a sweeping material prosperity.

d. Given what has been said about the previous problems, it follows that the question of *responsibility* within the context of technology presents great difficulties. That is true for those involved in the production process as well as for those in design. Because of the scientifically defined task, the worker, to the extent she still has a part to play in production, must completely disengage her freedom and responsibility. Or better put: these must *be* eliminated.

The responsibilities of the engineer, given how technology affects culture, have become heavier. At the same time she finds, possibly due to her one-sided background, but especially because of the situation in the workplace, significant obstacles in carrying out that responsibility. In addition, due to the intricate relations of cooperation, responsibility has more and more become a matter of ever increasing *communal* responsibility. We become even more conscious of the extent of this responsibility when we realize that we are often confronted with new technologies with which we have no experience. We do not have the advantage of being able to draw any lessons from the past so as to avoid old and new calamities. For example, many have the sinking sense that while recombinant DNA technology (see §19) will proffer unforeseen possibilities, it will also bring with it new hazards. Ignorance, innocence, a lack of experience, and increased risks go hand in hand (Jonas, 1984).

Clearly, our responsibility has become *greater* and *more difficult* to bear. We can give a number of reasons for this. First of all, we are uncertain with respect to almost all sectors of technological development—precisely *how* we can carry out our responsibility when it comes to modern technology? The operative range of the engineer, as we saw, is expanding, while at the same time the nature of her work keeps her from a comprehensive purview. Of increasingly little she knows increasingly more. She will have to submit herself in trust to her cohorts who work in an equally specialized manner. But it is precisely this specialization that weakens the community being called to bear a greater responsibility. Moreover, even where there is sufficient insight to bear the personal and communal responsibility, these efforts often are being frustrated by imposing scientific, technological, organizational, economic, or political allegiances. It is very difficult to break through an existing, dynamic trend, especially when it pervades the present culture.

e. The problematic situation with respect to responsibility deepens with the realization that the perplexities of modern technology are not incidental problems with one or more components, but that they all hang together. That is why partial solutions are also inadequate. An integral approach is required. But that in turn presents additional problems because our "technological culture" is at the same time a *spiritually disintegrated culture*. The Spanish philosopher of culture Ortega y Gasset put it succinctly when he wrote: "The world in which we live is the most technological there ever was, but spiritually also the most empty." There is no longer a unified vision of what it means to be human, of history, culture, and the future. The answer to the question of responsibility requires a communal response to the question of meaning and a communal sense of normativity. And that in no way is present any longer. That is why Jonas rightly speaks of an *ethical vacuum*.

8. Philosophical Perspectives

Both the advantages and disadvantages of modern technology have given cause for philosophical reflection, especially when considered in light of the development of the sciences. These reflections, however, have historically been highly divergent. Since the rise of modern technology and its problems, Western philosophers wrestling with the background of science and technology increasingly disagree.

Surveying that philosophical reflection reveals the spiritual abyss of our scientific-technological culture. Such an overview can also make clear that the problems we have mentioned have a deep spiritual-historical background and usually have to do with overestimating the potential of science with respect to technology. On the one hand, we find philosophers who evaluate modern technology positively. Other philosophers reject modern technology, seeing a growing conflict between humankind and modern technology. While the one takes modern technology to be an affirmation of human power and supremacy and of cultural progress through the advance of scientific-technological control, the other sees people alienated from nature and their own freedom as they become the victims of science and modern technology. In a word, a number of philosophers see modern technology in combination with science as a meaningful avenue towards a future that will the liberate us from many evils. Others see in this combination of science and technology a much more somber picture.

There are, nevertheless, so many shades of nuance within both categories as well as points of similarity between both manners of philosophizing that further elaboration is required. The following brief survey of philosophical viewpoints will show how central the human role is in these perspectives and how fundamental and significant one's view of science and its relation to technology is.

We will take a look at six different philosophical traditions: a positivistic-pragmatic position, a Marxistic one, a systems analysis approach, existentialism, neo-Marxism, and a counter-cultural view. Given the context it is not possible to be exhaustive. We limit ourselves to the main points.

a. The *positivistic-pragmatic view* of modern technology is very prevalent. This philosophical attitude is more often than not present in the minds and actions of natural scientists and engineers. Pronounced representatives are Karl Steinbuch and Norbert Wiener (Schuurman, 1980), both of whom were strongly influenced by the American pragmatism of William James. Their estimation of natural science and modern technology is very positive. Technology for them is the engine of progress in culture; growth in scientific knowledge is its fuel. They have put their hope especially in the development of the computer.

In the past, the so-called scientific design method (see §12) proved sensible and successful in controlling "dead" matter. If we are going to continue, extend, and strengthen this successful progress, so they argue, we are also going to have to use this method, as a method of control, for areas that are not specifically technological. We must analyze and form humankind as well as society and its future by means of this method.

When problems arise along the way, people blindly appeal to science, be it an established discipline or a new one, for the solution. They have their eye on a comprehensive, scientific-technological control of society. Such control can only be realized when economic and political forces cooperate. In other words, they favor a technocracy. They see anything that resists such an approach as possibly causing failure; it must be opposed. This includes all those philosophies that put any less trust in science, like existentialism for example. But they are equally opposed to religion, particularly the Christian religion. They argue that religion relativizes science and technology or sometimes even challenges their validity. For positivists and pragmatists, the closed, scientific-technological world picture is a *conditio sine qua non*.

b. The development of modern technology is also a central tenet of *orthodox Marxism*. For the Marxist Georg Klaus, for example (Schuurman, 1980), the person is pre-eminently the technological person who in and through the development of technology not only achieves full potential and freedom, but also undoes all oppression in achieving the goal of an emancipated society.

In contrast to the positivists and pragmatists, Marxists recognize that the development of technology goes hand in hand with many problems, threats, and dangers. Placing the development of technology in the hands of capitalists will result in the increased alienation and coercion of the labor force. At the same time, modern technology will bring an end to the class struggle. Marxists maintain that ongoing socio-economic development that proceeds on the basis of technological proliferation will wipe out the lengthening shadow of alienation and coercion and that an age of freedom will break forth with the dawning of the new day. In that day no one individual or class will have their hands on the reins of power, but humankind collectively will be the lord and master of the works of their hands. Along with the previous tradition, Marxists have a high regard for science as the means to get things under control. But because of their basic assumptions about the nature of society, how this will work itself out is quite different.

Marxists do *not* proceed from the freedom of individuals and from a free corporate means of production within which technology will come to fruition. Their assumption is that technology can facilitate revolution and ultimate emancipation only if it is orchestrated in a centralistic fashion and the use-value of the goods produced, over against their possible exchange-value, remains definitive. Which is to say, they proceed from the idea of a central technocracy.

c. The third group of thinkers are representatives, like Erwin Laszlo, of *systems analysis* (Schuurman, 1982). Systems science is about forty years old. This philosophy drew a good deal of attention because of the Club of Rome publications at the beginning of the 1970s. This Club of Rome made clear that present cultural developments in which science and technology play a dominant role are soon going to be confronted with over-whelming problems and perils. According to Laszlo, systems analysis is the only possible way both to analyze the dangers and problems of a growing world culture, with the use of models, and to turn these risks around and to resolve these problems.

Systems theory looks for the cause of cultural problems in an overly one-sided, hyper-specialistic, scientific approach to things. This philosophy is a proponent of a single science that will overarch all the separate sciences. Often a particular special science has a reductionistic influence on practical life because of the application or instrumental use of this science. Systems theory does not want to orient itself to one component, but to the whole of reality. For these philosophers the whole is more than the sum of the parts. And it is this whole that they are trying to get a bead on in their systems analysis approach.

An integral scientific approach via the system is meant to lead to integrated control. This kind of management is made possible by the newest of technological developments, like information and computer technology. From this we see that the system may not be equated with the *given* whole, but must be seen as an *artificial construction.*

Systems philosophy is trying to provide an answer to the cultural crisis. It is concerned about the dangers of technological development, about the degradation of nature, and about many other social, economic, and political problems. The assumption is that these problems have been elicited by the predominance of a *reductionistic* kind of science. This general systems theory is out to resolve the problems of what could be called first level science with a meta-level, more "holistic" science and with its correlate, a meta-level technology, namely, systems (or computer) technology.

The integrated control of reality that systems theory is aiming at, even on a worldwide scale, coincides with a fundamental shift. The prime objective is no longer the *progress* that positivists and pragmatists yearn for, but rather the *survival* of world culture. This major shift need not imply that the newest science and technology cannot be considered as guides toward a meaning filled future. On that score, we have to say with respect to systems science, that next to a change in orientation, we still are looking at a bolstering of the influence of (the newest) science and technology.

In passing, it should be noted that representatives of systems analysis have emphasized more than once that, on account of the central position of an eventual world government to control global culture and its future, Marxist and nonmarxist societies are approaching each other and will eventually work together.

d. Compared to the philosophical perspectives discussed above, *existentialists* are more aware of the problems of technological development (Schuurman, 1980). An important existentialist is Martin

Heidegger. He claims that humanity more and more is surrendering to technology and that in doing so is alienated from the origin of technology (Being) and from his own essence (freedom). "Technological culture" for him is one in which Being has been abandoned and forgotten. Though surrendering itself to technology and seeing to its advancement, humanity in fact finds itself on the road to madness. Other existentialists, like Karl Jaspers and Hermann J. Meyer, say that the development of technological culture threatens the human subject, particularly, attacking what makes the human personality unique, namely, freedom and individuality. They turned themselves against science and technology, which they see as autonomous, anonymous forces. It is not that they turn against the abnormal outgrowths of a technological society, but against science and technology as such. For them the scientific method operative in technology is a method of suppressive control.

Their protest is often impressive. But given their assumption that they are dealing with an autonomous, self-governing originating force, one can only ask whether they can provide a meaningful perspective for the future. They resign themselves to the necessity of the present situation, have a longing for the past, or want to flee or transcend contemporary culture by seeking refuge in a freedom beyond science and technology that is itself perpetually vulnerable. Through thought they try to rise above the thinking of scientific-technological control and to enter freedom's space of internality. This turn to the interior must repeatedly counter the ever present and proliferate domination of externality.

e. There is a good deal of similarity between *neo-Marxist revolutionaries*, like Herbert Marcuse (Schuurman, 1982), and existentialists in their evaluation of the position of man in contemporary society. However, they do not consider science and technology to be an autonomous power. Science and its relation to technology are not discussed in a negative manner, at least not primarily. Their critique is especially directed to economic and political powerlords, to an elite that takes science and technology into its service.

These neo-Marxists are inspired by an alternative vision for the future, a utopia in which everyone is free and happy, and try in a revolutionary fashion to transform today's society according to the model of that utopia. A permanent conversion of society will increasingly liberate it from the bondage of this present darkness.

The realization of their revolutionary ideas has extensive consequences for the development of science and technology and for the significance of that science and technology within economics and politics. Achieving practical *goals*, always in a domination-free manner (Habermas, 1970), must precede the resolving of practical *problems*. They rebel against the ideology of technocracy, to help human beings become authentic in daily life. In practice human beings will be less the worker and more the player. In that way there will be more room to satisfy what being human means, to enjoy life and to live out vital urges that now all too often are being suppressed.

f. Thinkers of the *counterculture*, like Theodor Roszak and Charles Reich (Roszak, 1969; Schuurman, 1980), argue as do the existentialists for the renewal of authentic thinking. They do not want to leave it at that, however. Authentic thinking must be confirmed in culture. Doing so need not imply the ways and means of the revolutionaries; who have a societal revolution in mind, after which also the way in which science and technology function will have to change. Representatives of the counterculture favor a spiritual revolution, an inner revolution of human consciousness. The results of such a reorientation would then undermine the present social situation and give rise to a new culture, the counterculture, characterized by an other science and technology. As do proponents of general systems theory, counterculture thinkers oppose reductionistic science and its universal application. Their preference, however, is not a reinforced or extended scientific-technological world picture, but its fundamental alteration. They want to return to a sense of the sacred, to the visionary consciousness that rests not in reason, but in feeling. They wish to be liberated from the total alienation of a scientific-technological culture and wish to attend to the nonintellectual faculties of the human personality, for abilities that refresh themselves on the splendor of visions. On the path back from expertise towards wisdom they are seeking for the good, the true, and the beautiful.

The values and the norms of the counterculture clearly differ from those on which western culture has been based since the scientific revolution of the seventeenth century. Cultural progress is resisted. Science and technology must be "adapted" so that survival once again is possible. The counterculture will have to be characterized by a small, human scale. Instead of everything being standardized, culture will have to be variegated, also in its societal forms. It will have to be more organic than mechanical; simplicity and frugality must replace profusion; varied, joyful work will have to come in the place of labor that alienates and is evaluated only in terms of productivity.

These suggestions of countercultural thinkers have greatly influenced many protest movements that oppose large scale modern technology (Schumacher, 1973 and 1977). But a consistent elaboration of these suggestions has not been forthcoming. As is often the case, countermovements often linger on as parasites to the present situation, all the while expressing a basic discontent.

9. The Knot of Science and Technology

When we survey the various philosophical views of modern culture, which were themselves shaped by modern technology, we conclude that many experience modern technology as very problematic. In addition, there is little consensus among the diagnoses of the cultural crisis, let alone on the therapy needed to resolve this crisis. Clearly modern technology raises many problems rooted in basic motives, or background forces, that bond science and technology.

In connection with the configuration of technology and science that gave rise to *modern* technology, it would be good to look at the general parameters of science's influence on technology. The French philosopher Jacques Ellul can help us here. In two of his publications, *The Technological Society* (1964) and *The Technological System* (1980), he analyzes modern technology culturally and philosophically. Some of the things he says come close to the position of the existentialists discussed above. What is worth noting and also very interesting is that Ellul also analyzes the characteristics of modern technology—features that can only be understood in light of the influence of science on technology. He makes the point that its scientific basis and the scientific method (see §12) are the reasons why modern technology is clearly different from the technology of the craftsman. An overestimation of science's potential in technology also suggests why so many problems (see §7) arise in contemporary culture because of technology or along with it.

Ellul says that modern technology is characterized by rationality, artificiality, automation, self-reinforcement, monism, universality, and autonomy. We will briefly discuss each of these.

a. The *rationality* of science shapes modern technology in such a way that its influence on culture is defined by this rationality. There increasingly appears to be less room for fantasy, beauty, and the irrational, says Ellul. By systemizing tasks, dividing labor, and dictating standard criteria for products and their production, less and less room is left for spontaneity and personal creativity. As each new technology is introduced, the schema of logic is strengthened in culture. Technological development follows the ironhanded laws of logic.

b. While technology was once imbedded in nature, it has now brought forth a thoroughly abstract, *artificial* world. Modern technology exhibits an accumulation of technological means that together begin to dominate and even eliminate nature.

c. The *automatism* of technology implies that technology more or less steers itself. Once scientific discoveries have been made, the technological development they make possible seems to happen of itself. In spite of all our deliberations, modern technology is not to be swayed.

d. The *self-reinforcing* character of modern technology is readily apparent in this process. As technology allows the science of physics to penetrate deeper and deeper into matter, its influence on technology increases in turn. This is clear from recent developments in nuclear research and technology. This self-sustaining feature of technology also reaches to education, economics, agriculture, politics, administration, and so forth.

The influence of the computer serves as a prime example. According to Ellul, the self-proficient nature of technological development also implies that it cannot be reversed. This irreversibility holds for technological development as such as well as for the influence that technology has on culture in its entirety. A new technological development often has rippling effects on all branches of technology and every sector of culture.

e. According to Ellul, one result of the automatism and self-reinforcing character of technology is that people have less and less to say about technological development. The phenomenon of technology, though blind to the future, nevertheless increases in intensity. People are affected by technology more and more, but in the process have less and less effect on it. That is what Ellul calls the *monism* of technological development. Weighing the advantages of technological development against the disadvantages of technological development is out of the question. There is also no place for moral considerations. Modern technology becomes totally separated from ethical deliberations. In fact, it is more often the case that ethics is being forced to conform to technology. Now that technology has become a system, the ethics that fits the need is a systems ethics, which does not critique technology as a system, but is there to support it as such. This does not mean, Ellul cautions, that technological development has put us on the road to certain catastrophe. The inner logic of technology, it seems, sees to it that the harmful effects are expunged. In other words, given the influence of science on technology, technology is in a position to correct itself.

This orderliness of technology in Ellul's estimation is devastating for human freedom. Technological order, he claims in typical existentialist fashion, eliminates human freedom.

f. Ellul sees the connection between the self-reinforcement and the monism of technology in the *plan*, taken in the sense of a definitive technological construct. What he especially has in mind here is another influence that science exerts on technology, namely, *universality*. By means of trade, wars, technological assistance, and the influence of telecommunication networks, technology's sweep spans the globe. Technology is disintegrating the barriers that exist between cultures.

The universality of technology is not just geographic. Ellul also has in mind a qualitative universality. Science focuses on what is universal, on what is true everywhere. In providing modern technology with a basis and a method, technology too has become universal in a qualitative sense. That is to say, technology, mirroring science's influence on it, is in turn imposing a uniformity on culture.

g. All the characteristics mentioned above can be traced back to the influence of science on technology. They find their unity ultimately in the *autonomy* of technology. For Ellul this means that technology becomes a world of its own, subordinating everything to itself and putting its mark on it. This all-pervasive technology will become so influential that people will subject themselves to it with the reverence of a religion. Not only ethics, but even religion will be influenced by technology and even defined by it. The society of the future is the technological society. The ethics of the future will follow the dictates of systems analysis. The religion of the future is a technological eschatology. People trust, admire, and worship technology for these kinds of reasons. But they will also come to fear technology's agents as though they were gods. The technological society is constructed by technology, for technology, and is exclusively technological.

As an existentialist, Ellul's critique concentrates on these characteristics of technology. But as far as the influence that science exerts on technology is concerned, certainly where that influence is unwarranted, it cannot be denied that Ellul's characterizations hit the nail on the head. In his later work (1980) as well he stays by this analysis. There Ellul avails himself of the general systems approach, which assumes that the whole is more than the sum of the parts, and analyzes the phenomenon of technology more as an all-inclusive, coalescent system. Technology's totalitarian traits are apparent from its being a system in which all the technological means are incorporated into a common context.

In Chapter 3 we will consider the ethical questions surrounding modern technology. We will see there, in contrast to Ellul's claim that ethics is banished, that because of the advancing influence of science on technology and its consequences, ethics is being given more attention. The big problem with respect to modern technology is that the influence of science is something people now take for granted. In what follows we will see that the characteristics of science, for example, its universality and abstraction(s), are in tension with the fullness of reality. In reflecting on these matters ethically, the central question will be whether technological science, instead of dominating technology, can also seek to serve that technology.

The next chapter analyzes the structure of the relationship between science and technology to discover what makes modern technology unique. By taking a closer look, particularly at the scientific method of which technology avails itself, we will be in a position in the fourth chapter to deal legitimately with the philosophical-ethical questions and problems surrounding technology and the technological sciences. In the end, it is only within that context that we can perceive all of culture meaningfully.

Chapter Two

TECHNOLOGICAL SCIENCES

10. The Foundational Sciences

The immediate foundations of modern technology are the technical sciences. Given the history and differentiation of technology, from materials and energy technology to information and biotechnology, there is sufficient reason to speak about technological sciences in the plural. In what follows, however, I will refer simply to "technological science" and mean with this term the wide diversity of technological sciences (see §18). Each of these technological sciences reflects the scientific knowledge of the respective sectors of modern technology.

Technological science has to do with knowledge about technology, taken here in the plain sense of forming and making. It should be noted that there exists more scientific knowledge about what is formed and made (e.g., technological things and processes) than what is here referred to as technological scientific knowledge. An obvious example to illustrate this would be physics. In technology we try, for example, to give form to inorganic or "dead" materials like stone or wood. Scientific knowledge about these things from physics serves as a basis for technological science.

The distinction between natural science and technology, as theory on the one hand and technological praxis on the other, loses its sharpness because technology is being defined more and more by the natural sciences. Many technological things and processes have become so complex that only scientific knowledge can provide us with insight about them. It is becoming increasingly clear that technological science bridges the gap between natural science and technology.

Physics and technological science are closely related, as is readily apparent when one compares these sciences. Why closely related? Physics seeks knowledge about the physical facets of things, while in technology physical subjects are the starting point for technological formation. As a result, technological scientists must know about the physical features of things before they design anything.

To the extent that humans are no longer needed in technological execution, physics and the science of technology look alike. Terminology in both sciences often sounds the same, though the notation in technological science meets the needs of the concrete technological scene. In other words, in technological science one prefers to choose symbols that conform with (automated) technological formation or execution. Drawings, graphics, diagrams, nomograms, functions, and descriptions, depict the technological order. Symbols in physics, on the other hand, are more abstract because they are directed to the physical structure of things.

The points of similarity increase the more technological science deals with fundamentals. Similarities decrease the closer technological science gets to technological practice, because there it primarily prescribes. Some examples are designing a bridge that will require less material, building a dam that is safer, or devising faster train travel. When technological science is so involved with concrete technological practice, it is also more limited. Scientific-technological research ends as soon as a solution has been found. Research can begin again if there is a desire to improve products or if socio-economic circumstances change, or if improvements in a related technology make new developments possible. But it can also be that development in a technological science comes to a halt because no new technological possibilities appear on the scene. So, the more technological science deals with practical matters, the more it differs from physics (see §18).

One should also see how physics and basic technological science differ. Physics has to do with knowledge about nature. Technological science is based on general knowledge of how technology is formed and how it uses physical subjects. This is specific knowledge, namely, knowledge about forming technology.

Physics has a more universal focus than does technological science. Physics is after universal or generic knowledge about the physical dimension of reality. The science of technology attends more to individualized knowledge about the technological formation of physical things and processes. From this point of view mathematical physics belongs to the basis of technological science.

Finally, concerning their connection and difference, one ought to keep in mind that physics in its pursuit of knowledge increasingly uses technology and the insights that accrue to technological science, for example, through experimentation. Physics, technological science, and technology interact (Van Melsen, 1961; Schuurman, 1980). On the one hand, physics, together with other foundational sciences, lays the foundation for technology, while on the other, technology and insights gleaned from technological science create the possibility for further development in the foundational sciences. An example of this would be when observations and experiments in which answers to questions concerning scientific knowledge of the physical facet of reality can be gained with the aid of technological mechanisms. In that case the technological apparatus serves the process of theory formation in physics. This technological support can lead to new knowledge in physics that in turn can enhance technological development. A scientific discovery can be the condition or catalyst for an invention, because with a scientific discovery the frame of reference within which invention occurs has been expanded. The invention can in turn benefit the further development of physics (Van Melsen, 1961; see §14).

That technology permeates physics is quite obvious from the apparatus used in research laboratories. But it is also clear that the most advanced of instruments, for example the particle accelerator, result from the newest scientific developments. And yet the distinction remains a clear one: natural science describes the material world. It does not change that world. In research, the technological apparatus serves knowledge; it is not itself the goal of knowledge.

The scientific basis of technology and technological science, as was said above, is not only physics. Because what is known in physics is increasingly assimilated in the form of mathematics, we often deal with mathematical physics. In addition, this knowledge has most recently been modeled according to the rules of mathematical, information theory, so that information science too must count as a foundational science (see §6). Other foundational sciences, besides mathematical physics, are biophysics, chemistry, biochemistry, and, to the extent we are dealing with biotechnology, also biology (see §18).

We will see later that biology and ecology should actually also be included with the foundational sciences because technology always takes place in an environment interrelated with the living natural world. We live in an age of environmental pollution. We see nature degraded and even destroyed. We see that the technical schools often include only those foundational sciences that help us continue to expand technological functioning, without regarding its consequences for nature and the environment. Fortunately many are beginning to realize that the damage this has caused must now be undone and as much as possible be prevented in the future.

Because technology functions significantly in society today, and because, for example, we must insure the quality of work, it is a good thing that in the last twenty years the social sciences have been accepted as the context within which technological science should be pursued. In this study, however, we will not be able to discuss how important this is for the development of technology.

11. Development of the Technological Sciences

We have mentioned in passing that in modern technology, *design* has disengaged itself in place and time from *implementation.* The connection between both of these sectors of modern technology is formed by the objective construct or design. It results from productive imagination and functions as the prescription for the implementation. The new construct provides the criterion for the technological execution. As knowledge about that technological process, it belongs to the field of technological science.

Because preparation and execution are usually separate in modern technology, these two areas can also be distinguished within technological science. On the one hand, based on present technological development, we know about the technological process. On the other hand, part of technological science involves knowing about how engineers go about designing. This knowledge in turn leads to results having to do with the first sector. Seen in this way, the engineer is the architect and builder of technological science. He is both the scientist and the technician.

Because of the computer, the first sector—actually implementing and forming—will become completely automated. Human beings will actively participate less and less in production.

In the second sector of technological science, the engineer plays the major role. He does not simply follow his intuition or proceed on past personal experience. Instead, given his scientific knowledge (see §10), the engineer distances himself in a typically scientific manner from the immediate environment within which the technological problem has arisen. Design is based on science and is characterized by a scientific attitude or method.

The very first thing an engineer has to do is state the technological problem. Then he asks whether what technological science knows at present provides sufficient grounds for him to resolve one or more of these problems. Usually it does not. So the engineer proceeds from scientific knowledge to analyze the problem, abstracting and isolating it from the immediate context, breaking it down into modules or unit problems, and then posing it in general or universal terms. This is what we call the *scientific design method* (see §12). The engineer, using his technological imagination, tries to solve the subproblems. To solve a concrete technological problem, then, he or she coordinates the various module solutions and integrates them into a whole: *the design*.

While designing, the engineer must check repeatedly to see whether the design is suitable to the technological problem and to the context in which that problem arose. With complicated problems this feedback is especially imperative. Technological knowledge and imagination alone cannot produce a good design. Performing experiments or making prototypes is required to answer questions that technological science cannot (yet) resolve (see §17).

The ultimate intent of technological science, in contrast to physics, is not to collect knowledge, but to develop technology. We could also say that the science of physics goes its own way, while technological science always follows a detour to promote technological development. Physics continues to aim at universal knowledge, while the goal of technological science is new technology. The hypothesis, anticipating a future physical theory, is for that reason analogous to the innovation or invention anticipating a new design (see §14). Here lies the similarity. The difference is found in the direction both sciences are headed (see Figure 1).

Science	Technology
existing knowledge	existing product or process
│	│
via hypothesis and reflection	via recognition of needs or market research
⇩	⇩
hypothesis	new innovation or invention
│	│
via logic and mathematics	via feasibility study (technological science)
⇩	⇩
falsifiable deductions	adequacy of the design; testing
│	│
via experiment	via prototype/development
⇩	⇩
confirmation	production
│	│
via communication	via public acceptance
⇩	⇩
new knowledge	new product or process

Figure 1 Development in science and technology (see Rogers,1983)

A very common problem for engineers today is that the technological knowledge of the individual engineer does not increase proportionately with the tempo of technological development and hence always lags behind the breadth of relevant knowledge. Specialization and teamwork are supposed to alleviate the effects of this problem. Through this specialization the engineer increasingly loses contact with the context in modern technology. Specialization can weaken technological development. Working in teams is one attempt to prevent this. This encumbers the task of the engineer, but the advantage is that more can be achieved in a technological sense.

A special point deserves attention here. In contrast to the foundational sciences we have discussed and classified with the natural sciences (see §10), technological science is a *social science* because it tries to formulate knowledge about *human activity* in a technological sense. The free and creative human being plays a decisive role in its field of investigation, that is, in the entire arena of technology. This is especially true when it comes to designing and inventing, but also to the extent humans have a place that as yet cannot be automated in the process of technological production. If automation becomes fully realized, this would mean that the limits for the technological scientific control of the production process will hold as the limits for a particular field within technological science (see §12). In separate paragraphs along the way we will analyze technological science as a social science and the special problems this presents.

After scrutinizing the scientific design method and its limitations we will turn to invention. Inventing breaks open this method because in and through the process of inventing we find solutions hitherto unknown. Inventions can obviously differ in kind. What they have in common, however, is that they always bring with them new possibilities for the development of technology. But they often also bring new problems.

12. The Scientific Design Method

As was said, the engineer uses the scientific method. This method is like that of the natural sciences, but is technological because it focuses on the design of technological things or processes. This scientific design method characterizes modern technology. As a result, the features or characteristics of science are projected in technology. Seeing this close intertwinement of technology and science also helps us see the characteristics of modern technology. In the next paragraph we will see that there are limits to this method and that understanding these limits also helps us see the limits of modern technology. But we will begin by investigating the scientific method.

With the scientific method of the natural sciences we attempt to discover knowledge of reality. With the scientific design method we intend to formulate technological designs both for the product and for producing it. The ultimate aim is *a temporally and spatially remote, scientific control* of the technological process of forming or producing. But its result also belongs to the domain of technological science because the designs furnish us with knowledge regarding technology.

Some of the characteristics of natural scientific knowledge and of its method are reflected in a technological kind of way in technological science. This is in large part due to the scientific basis of design. These characteristics can also be found in the technological production process and in its results.

Analysis, abstraction, and synthesis constitute the method of science. That is to say, in science we consciously turn away from and seek to avoid the influence of the immediacies of the concrete individual, situation, or environment. As a result, scientific knowledge is universal, enduring, coherent, and functional knowledge.

The engineer, too, follows the same procedure. The technical question is broken down into parts which in turn are posed as universal (sub)problems. Universal modular solutions are sought both for the components of the thing to be produced and for steps of the production process. The division of functions continues until one has arrived at atomized, isolated, autonomous functions.

Analysis puts technological modular functions at one's disposal. Abstraction presents each of these functions on its own, bracketing out the others, so as to find a technological solution for that subfunction. The result consists of basic building blocks that have a standardized neutrality of purpose about them and, hence, can have a universal application in technology, as is the case for example with propellers, rivets, or dovetail joints. In this universal applicability, a result of analysis and abstraction, we find science's general or universal knowledge projected in technology. In other words, the universal suitability of the solution of a subfunction is an analogy of scientific knowledge in technology.

In addition to the neutralizing division of functions, the scientific design method, as was said, also aims at a spatially and temporally remote control of the process. Therefore, the tool or technical operator must be constant over the course of time. The control of technological formation at a distance and over time is an analogy of science as enduring abstract knowledge of reality.

This remote control, with respect to the whole, is further made possible by the integration of the neutral modular solutions. Depending on the modules that are merged, this integration of functions can represent many forms of individuality. As such, it is a technological analogy of synthesis in science. *The scientific design method aims then, on the one hand, to standardize how functions are divided and, on the other, to anticipate how these functions will be integrated for various specific ends.* The production process that results is likewise an analogy of the *coherency* of scientific knowledge, with the product of mass production showing something akin to the *universal* character of scientific knowledge.

In like manner, an analogy of scientific abstraction can be noted in the integration of the solutions from each module. These problem sets were originally isolated by means of abstraction. After a standardized solution is found for each subset, the integration of these separate solutions requires that technological formation take place elsewhere, in isolation. Pertinent measures are taken with that in mind. Technological formation has an enduring character about it because influence from the surroundings, like temperature and humidity, are precluded as much as possible.

This scientific design method determines the basic structure of modern technology. Technological science and technology are virtually interwoven. An example can help to elucidate their relationship.

A technical operator, as we said, is a tool that works automatically. Given such an operator we can distinguish the thing and the process of the transformation of energy. We can note the influence of the method on this thing, the machine. The parts of such a machine have a general or universal purpose and are as such technologically neutral. Together they make up this particular, concrete machine. The neutral division of functions, then, is allied to an individualizing integration of functions.

The energy exchange that drives the machine has undergone the same effect. The flow of energy is in fact individually directed for this particular machine, for example, a lathe, but in previous steps of the production process is dependent on energy of a more universal structure, that is, one in which the energy flow must have as little individuality as possible. Universal neutrality in a technological sense is always accompanied by individualizing integration.

When developments in science lead to a new method, as was the case with systems analysis, then this method recurs analogously at the technological level. That is why general systems theory and computer technology are complementary.

The method of systems analysis (Laszlo, 1974) chooses not to proceed from analysis and abstraction, because that would fragment and reduce reality. Systems thinking attends to the whole of reality. According to Laszlo, that is why its method is expansionistic and synthetic.

Without going into great depth about this method, I must point out that it provides a new scientific prospect towards advancing the cause of modern technology. At the same time, one must be warned of a possible misunderstanding. Although the systems analysis approach differs, it is nonetheless a scientific method, with the same promise of possibilities and potential for problems for technology that come with scientific methodologies. Its possibilities and problems are simply different from those of the method of analysis and abstraction.

The most striking difference lies in the shift in scientific focus. Whereas earlier attention was given to a component, an aspect, or a function of reality, in systems theory the *whole* remains central *as a system. That whole is more than the sum of the parts.* Closer consideration shows the whole to be the sum of the parts plus the interaction between those parts. The interaction is controlled by additional information, communication, feedback, equifinality, self-stabilization, and self-organization. This approach also attends to the system and its environment in terms of the input and output of material, energy, and information. We see from this that although the method is new when compared to previous methods, the result leads to an enhanced continuation of modern technology in information technology, computer technology, and the technology of integrated systems.

13. Limits to the Scientific Design Method

That the scientific design method we are describing is technological, and not something of the natural sciences, appears from the fact that reality resists the realization of this method. Owing to its scientific character this method has a universal pretense. But because it is also directed to a concrete technological application, this method must conclude with individual designs or techniques. That is why applying this method in the practice of technology has its limits. Its possibilities are in principle limited because science is not *congruent* with concrete reality. The universal, abstract, and enduring or resolute character of science pulls in a direction other than the fundamental features of reality, where everything is singular, hangs together with everything else, and changes. People try to suppress these features of reality with the scientific-technological method, but they are never entirely successful. Concrete reality, in which technological forming takes place, sets limits to the scientific design method. For example, entirely universal products cannot be made, the technological process can also not be fully isolated from its environment, and finally, technological things that are altogether resistant to wear are out of the question. It is the resistance of reality that prevents an all-encompassing scientific control of modern technological formation. A scientifically recalcitrant remnant of reality will always, in one way or another, continue to make its demands and to confront the realization of remotely controlled, total automation with singular problems.

The resistance of reality manifests itself in many ways: in occurrent variations in natural materials and energy sources; in perpetually changing factors in the technological milieu, like temperature, humidity, pressure, and so forth; in changes in the technological operator due to wear, corrosion, aging, breakage, fatigue, fluctuating temperature, and the like; in the individuality and freedom of the workers, to the extent they must still necessarily be incorporated in the technological process of formation; and in the fluctuating tastes of the consumers of the products to be made.

These variances present problems when we try to control things scientifically from a distance. However, these problems are not always insurmountable. With the introduction of the computer, for example, digital control has made great strides in satisfying the individual desires and wishes of the consumer while maintaining remote control (i.e., complete automation).

Although a complete scientific formulation of technological formation is not possible due to these variances, it is interesting that this does not always mean that remote control is out of the question. We see here the fruit of general systems theory. Certain degrees of variation owing to natural inconsistencies can be corrected with the help of a "trick." At the point where a set standard is not met, an automatic switch is triggered and on the basis of the cybernetic principle of feedback, the deviation can be nullified. The occurrence and frequency of these variances cannot be theoretically predicted.

Such limits are due to the laws of nature. The success of the scientific design method easily tempts people to absolutize this method and to apply it where it ought not be used. They forget that technological science is a social science. The relevance of this design method is overestimated when, for example, the worker in the production process becomes an object of scientific-technological control. The situation becomes very serious when the scientific flow-chart dictates that, due to a lack of technological operators, workers have to repeat monotonously a single subfunction. The individuality and freedom of workers has been eliminated because these disturb the degree of scientific control. They are expected to work as mechanical parts.

A good deal of labor's drudgery, for example, the tedium of work on the assembly line, has disappeared with the increased application of the micro-computer. But that has often brought with it the problem of unemployment. The complete automation of the production process requires highly qualified workers. Each step of the process once performed by human beings is now projected or objectified in the technological operator. But each of its steps, whether that used to be head or hand work, expediting or controlling, can after all only be worthwhile if there is, at its side, a person who remains physically and mentally the master of its work (Van Melsen, 1961).

Overestimation of the scientific design method is also often evident in mass production. Mass production is not so bad for making component parts. But when the final mass-produced product is supposed to serve the consumer, many problems surface. Van Melsen rightly points out that a mass-produced product can have an unmistaken leveling effect on the user. The statistically average end user, as the cross section of many users with a diversity of choices and desires, ultimately becomes the actual user. Every consumer eventually becomes a representative of the average user. People begin to adapt themselves to what technology prescribes: standardized goods for standardized users, that is, for the masses. In all honesty, we have to say that these mass-produced wares are cheap and are available to many who otherwise would have to do without.

The objections and dangers are greater when the scientific method becomes the absolute in technology and is given regal status (see §21 and what we will call "technicism"). When "technologize" is the answer to every question and "technologically possible" dictates the agenda for tomorrow, the limits of the scientific design method are being or will soon be transgressed. It was especially this absolutization of the scientific design method as well as its imperialism that Ellul had in mind (see §9) and that constitutes a problem for the ethics of technological science (see §§21 and 22).

14. Invention

The pivotal point for human work in technology, then, will be in the design phase, in preparing for the process. To that end, people will need to learn the foundational sciences of technology, technological science, and the scientific design method.

Even though designing is dominated by this method and it guarantees the continuity in technological development, we must attend to breaks in this continuity. We are talking here about *inventions.* Technology continues to develop in the context of scientific consultation and formulation. In ways that are often unexpected and cannot be simulated, productive imagination can put people on the track of technological innovation. These innovations or inventions can interrupt the developmental process of technology.

Where invention begins is difficult to ascertain. Usually a number of factors like need, coincidence, want, playfulness, imagination, together or individually, play a role. But that is not all. The unthought-of dimension, which makes invention what it is, is not sufficiently explained by these factors.

In modern technology, invention takes place at a higher level than in classical technology. Today inventions build on mathematics, physics, information theory, and technological science. This also means that when inventors are onto something new in technology, they try to develop it scientifically and to check it out with precision instruments to complete the invention. But at the decisive moment of inventing, theory plays no role. The invention and its effects will later belong to the field of technological science, which also served as its basis. This being so, the invention contributes an additional path to those presently available for new technological possibilities.

Among inventions we can distinguish primary inventions, which have to do with an entirely new technological unit, and inventions that improve on components. Dessauer speaks in this context of "pioneer inventions" and "developmental inventions" (1956). In this connection, however, we should add that because it is becoming very difficult for one person to invent something new, the inventor more and more must work with a team of innovators.

It is important to note that although technological science affords a more advanced basis or context for invention, there is a danger that this basis for invention or innovation can stifle the development of modern technology. The sheer growth of theoretical means and their ready access threatens to sterilize the productive imagination. On the other hand, invention can be a opportunity to break through an unfavorable or infertile development and to open up new perspectives. This is especially true in developing computers, where even playing around can spark new inventions.

Raby was correct in pointing out that the technological realization of inventions often depends not only on the scientific basis, but also on the material capability of technology itself. Already in 1823 Babbage had worked out a calculator in theory, that is, on paper, but its realization met with many difficulties. Apart from financial constraints, the technical ability was not yet available.

Remarkably, the reverse situation presents itself at the moment in computer technology. Computers offer many more possibilities than there are applications—the prospects precede the novel technological ideas. The human imagination is at the moment not in a position to oversee and take advantage of all the possibilities provided by computer technology.

Inventions in technology spring from "objective" elements as well. (The same can be said analogously of discoveries in the sciences.) More than once inventions have occurred simultaneously in different places. Stork speaks in this context of the "births of time" (1977). Given the scientific foundation of modern technology we see this phenomenon more in our day than previously. Because of the scientific basis, the phases of preparing, incubation, illumination, and verifying or testing, have become clearly distinct and inseparably connected with the process of invention. Though imposing to the individual inventor, they at the same time actually invite many to become inventors. This situation gives rise to the idea that inventions can be more or less planned. That is why research laboratories try to program for future inventions and innovations.

Once inventions become known, variations are quick to follow. In connection with patents, it is important to note the legal requirements placed on an invention. They must include a new, precisely defined, distinct, and integral assembly of technological things or processes. Inventors must orient themselves in two directions. First of all, they have to determine what is and what is not included in the invention. What defines its individuality? What makes it peculiar? Secondly, moving in the opposite direction, they have to answer the question whether their invention is not a specialized application of a more general order of technology that made this new invention possible.

In summary, inventing, because of human creativity, cannot be fully analyzed scientifically. Psychological studies, however, do show that creativity, as an expression of human freedom, can be nurtured with special education, through "brainstorming," and by learning to deal with critique. These three completely coincide with the following traits of an inventor: flexible thinking; a playful, young, inquisitive mind; the ability to see what everyone else does differently and, hence, to suggest nontraditional solutions; a vibrant intuition; a desire to perfect a technique; a bit of an introvert; great intellectual prowess paired with courage and a reticent social attitude. These are all conditional factors for invention. When it comes right down to it, however, inventing eludes scientific inquiry (Schuurman, 1980).

Finally, modern inventions are founded on science. Inventions can form a prologue to the rise of completely new scientific knowledge. At the same time, scientific discoveries can likewise be the catalyst for new inventions. As these things happen science and technology interact and the complexity in technological science increases, requiring a clear analysis and definition of place.

In previous paragraphs we saw that the scientific design method is an analogy of the natural scientific method. Inventions and innovations can be compared with discoveries and hypotheses informing scientific theories. (See Figure 1 on page 43.)

15. Technological Science as Social Science

Technological science encompasses any scientific knowledge that has to do with designing, inventing, and technological execution. In other words, technological science also investigates human beings. That is why technological science, in contrast to physics for example, is a *social science.*

Technological science, as a social science, is the first science in the transition from the natural sciences to the social sciences. That is why there is much in technological science that is similar to the natural sciences. We can see this from another angle as well. Technology, we said, is the tool-aided formation of the natural side of reality. The preparation for and execution of technology belong as well to technological science's field of investigation. When it focuses on fully automated technology in the production process, technological science is dealing with a determinate reality. Obviously, similarities with the natural sciences will be the greatest in that part of technological science. To the extent it deals with the engineer as designer and inventor, the weight will fall much more on technological science as a social science.

People quite rightly speak of technological science as a *normative science.* The engineer and the inventor are obliged to seek normative principles for technology, especially for how technology should relate to human beings and to nature. A clear example of this, mentioned already in §13, is that in a production process that is not fully automated, workers may not be dealt with as though they were technological objects and be expected to work as though they were components of a machine. When this happens in technological science, human freedom, creativity, and responsibility are undercut. That is why technological science, as a social science, must give heed to the quality of the work that the process requires as it seeks to combat and circumvent drudgery. The engineer and inventor must also follow normative principles when developing and renovating technology. Designating a direction for technology is a normed, cultural business (see §24).

We find then in technological science two distinct sectors. Producing and executing, whether it is fully automated or not, is the first sector. The second sector has to do with the designing, inventing, theorizing, and experimenting that accompany these (see §17). There is interaction between both sectors. The design sector has to solve problems that arise in technological production. These solutions at the level of technological science serve, in turn, to instruct us in forming technology.

This scientific control of the production process has its limits. The surroundings in which production takes place cannot be scientifically determined in their entirety. The same holds for the aging and wear to which the tools, or technological operators, used in production are subject. Products and the raw materials used to make them also display idiosyncrasies that can be factored in and controlled beforehand only within certain limits. In addition, at each point along the way there is always the possibility of mistakes or breakdown due to severe, unexpected irregularities. If these difficulties are not covered by automatic adjustments or switches, remote scientific control has to leave room for the responsibility of the workers in the production process. This kind of responsibility also appears in the final stage of manufacturing an individual product; although digital computer monitoring does offer possibilities here.

In general, we can say that the influence of technological science is such that people will want as much as possible to automate production or technological formation according to a scientific design. The inherent inclination here will be to rid it of the human presence as much as possible. Consequently, any attention within technology for human beings in their freedom and responsibility has come to focus more and more on the design phase, although there too scientific operations can be automated with the help of the computer (see §17).

16. Problems in Technological Science

The development phase of technological science is defined by a number of constraints. The problems with which the engineer is confronted arise in a specific context and are defined by historical factors and, further, by industrial conditions. The immediate context is determined in a historical sense by the amount of reliable technological knowledge about production processes and products fabricated to date.

Technological science is also conditioned by the actual state of technology within a particular industrial plant.

A separate problem is that technological knowledge grows by leaps and bounds and few designers can keep up with all the developments in technological science. This has become one of the principal problems in modern technology. The obvious solution to this problem is to step up specialization. But this, too, has its limits. The more specialized people are, the less they know about life beyond their specialty. That is why the dangers of specialization are offset with teamwork. For the rest, steps must be taken to make the newest developments in technological science available to designers in a way that is both quick and efficient.

Although technological problems arise in a defined, particular context, it is important for the technological scientist to be able to pose and to resolve these problems in a way that is as general or universal as possible. This can result, however, in a tension between the universal solution and the individual context in which the problem surfaced. To the extent the outcome varies from what the original problem required, the original context will have to adjust itself to fit the solution. The issue here is once again the tendency to neutralize individuality and the functional integration that accompanies it.

On the other hand, we have to keep in mind that the scientific treatment of technological problems often arrives at solutions that have a wide-spread significance. Actually, one finds a broad scale of possibilities between the extremes of the most universal and most individualized solution. Illustrations of the first category would be, for example, the transformation of energy from one kind into another (photoelectric, thermochemical, electro-mechanical, and so forth) and within one energy group (e.g., lenses bending light rays, the damping of energy, mechanical transmission, etc.). When people try to solve broad problems for all time, they tend to design neutral building blocks (like photo-electric cells, capacitors, gear wheels, and electron tubes) which, when integrated, as we saw, can lead to an individual final product.

Thus we see that technological science is busy with problems of universal consequence and with problems of individual specificity. The first category belongs to the foundational technological sciences, the second to the more practical or applied technological sciences. (See §17 for a survey of the technological sciences.) If engineers would limit themselves to the applied technological sciences, they are going to lose contact with the breadth of technology. That is also why students in technological science more and more feel frustrated when they lack a solid grasp of the technological sciences in general. At the same time they realize that their inability to take insights from elsewhere and constructively apply them directly to their own field is the price they will have to pay for this deficiency. Think, for example, of the generalization of vibration systems for the different kinds of energy processes. In addition to the specialist's knowledge of the applied technological sciences, students still need to know the foundational technological sciences.

From a scientific point of view specialization is understandable. But it has its drawbacks as far as the development of technology is concerned. The curriculum for scientific, technological training is going to have to strike a good balance between the foundational, scientific knowledge of technology and knowledge about the practical face of technology. This tension between universal and individualized scientific-technological knowledge may not be resolved on pragmatic grounds by emphasizing the latter. Universal knowledge about things present throughout technology, such as measurement, connectors and regulators, and energy transfer, is an indispensable prerequisite for applying this knowledge in individual cases. (See Figure 2 on page 65.)

17. Scientific Operations

The engineer appeals to technological science and its possibilities both when beginning to design as well as when finishing. At the start, design has to do with a scientific formulation of the technological problem. When finished, the design, as the law or order for the technological solution, is placed at the disposal of technological science. In the meantime, scientific operations are carried out in the designing that continually emphasizes the relation between designing and technological science.

The scientific operations are set up according to the scientific method, the purpose of which is to come to a symbolic objectification of the technological structure or order. We can distinguish two groups of operations: (a) *theoretical* or *deductive operations* and (b) *experimental* or *inductive operations.* Given their significance in technological science we will briefly discuss each of these operations.

(a) *Theoretical operations,* in the first place, make it possible to work with technological questions and problems. They also are aimed at deriving new designs for technological problems given the present state of the technological sciences. This usually involves moving from what holds in general to what holds for a particular, concrete, technological question.

The symbols used in these operations are usually borrowed from concrete technology. There are two types. Symbols that represent things denote the relative durability of technological operators or tools. The other symbols refer to changes in the technological processes. When scientists use arithmetic symbols, they refer to them respectively as *constants* and *variables.*

The symbols for constant and variable magnitudes can in turn relate to the physical, spatial, and numeric aspect of tools and processes. A few examples of symbols for constant magnitudes, that is, for the lasting characteristics of things, from the technologically unfolded physical aspect would be these: self-induction, resistance, mass, resonance, but also the elastic moduli, permeability, and the dielectric constant; from the spatial aspect: the (technological) moment of inertia and the cross section of a conductor; from the numeric aspect: the number of coils of a spool. Symbols for variable magnitudes for the physical aspect of technological processes are, for example, the strength of the current and the field, speed, and the amount of energy; for the spatial aspect, the volume of the cylinder between piston and head; for the numeric aspect, the frequency and revolutions per minute.

Among scientific operations we can distinguish the *argumentation,* the *depiction,* and the *calculation.* Each has its own function and as such, supplements each other. Their use depends on how straightforward or complicated the technological problems are. For a simple problem, calculation is often sufficient. For a more complicated one, argumentation or demonstration is called for. In general, argumentation is less effective, certainly than calculation, which is the most effective. One cannot, however, always use calculations. When people play an important role in forming or producing technology, one has to revert to diagrams and sometimes simply to a verbal rationale. On the other hand, it often happens that one begins with a verbal presentation. Argumentation can then, after necessarily discounting certain factors, give way to depiction and calculation. The limits of the scientific, technological method, discussed previously (§12), play a part in leaving these factors out of account. They have to do with the tension, also mentioned previously, between science and the individuality, change, and coherence of reality.

Related to this tension is the fact that calculating can be applied most easily at the most general level of technological science. The more the design is individualized, the more difficult calculations become and the more will depiction and argument take over.

A very prevalent approach, that again makes calculation meaningful once certain factors are disregarded, is determining the safety coefficient of stressed technological constructions. By methodically leaving things to the side as one proceeds further, the technological focus remains. As a result, these calculations may not be considered simply a matter of doing mathematics. An example may help. When calculating for a truss bridge one represents beam junctions as though they were hinges while in fact that is not the case. Because of this simplification one must assume a large safety margin so as to once again make the calculation possible.

Important types of theoretical, deductive operations include *calculations,* the graphic formulation and representation of the order or structure of technological things in *blueprints* and *drawings,* the graphic formulation and representation of the order or structure of processes in *diagrams* and *graphs,* and *technological instructions.* Theoretical operations can be applied in technological science just as they can in physics. The only difference is that in physics these operations have a universal scope, while in the technological sciences, which ultimately have to do with concrete technological things and processes, they will be more or less individualized.

Calculations have to do with relations between arithmetic symbols which stand for characteristics of technological magnitudes. One carries out mathematical operations on these relations, and in doing so, gains insight into the structure of technology, insight not gained through argumentation and demonstrative statements.

Graphics focus on the structure of technological things and leave the implicit energy processes out of the picture. The design is objectified in the form of geometric symbols. Arithmetic symbols and brief descriptions may supplement the drawings. The geometric operation takes place as the drawing is being made and usually goes hand in hand with calculations.

Graphs and diagrams, in contrast to drawings and schema, always depict a technological process.

Instructions can be either a description of a technological thing or technological process. As such, they connect drawings and schema, on the one hand, and graphs and diagrams, on the other.

(b) In technological science one might shift to *experimental operations* if theoretical ones, and particularly calculations, are no longer possible either because a good starting point is not available or they will not withstand the course of time. Chemical technology, on the other hand, which has been around for long time, cannot do without experimenting. Qualitative chemical analysis and classical organic chemistry are prime examples. In technological science, however, one usually experiments to confirm or to correct a theory.

An important characteristic of experimentation is that one arranges reality in such a way that one gets a clear answer to the question being posed. The answer is received via *instrument measurement* in the form of arithmetic or geometric symbols and then assimilated into knowledge about the technological structure of things and or processes.

When following the deductive method of theoretical operations, one usually proceeds from the general or universal and moves toward the particular or individual. The experimental method, however, requires that one move in the opposite direction, namely, inductively. One proceeds from the individual results of the experiment—though many individual features already have been eliminated by the arrangement of the experiment—with an eye to arriving at knowledge or a point of view that is as general or universal as possible. This tension is the same as occurs in theoretical operations. In that case, the procedure of methodically leaving things out of the picture is necessary because of the individuality, coherence, and change of technological reality that science cannot control. In experimenting one tries to eliminate these factors by concentrating more on the question the experiment is meant to answer. Despite all this, one will *not* find much in the final calculations that actually belongs to the technological structure, and all kinds of things will *implicitly* be incorporated in the results of an experiment, such as the influence of the environment and the interrelations of the measuring circuit, that were *not* intended in the question.

Both of these operations interact with and supplement the other. They can guide and stimulate the productive imagination of the designer, and at higher levels, help develop technology.

A special point deserves mention. Today scientists use computers in the operations of technological science. This is a prime example of how much science and technology have infiltrated each other. Computers speeded immensely the development of the technological sciences. The engineer can now do much more than ever before and in less time. Theoretical questions today can be so complicated and wide-ranging that engineers alone, either individually or even with cooperative effort, cannot solve them. The computer affords a remedy. The same is true, more or less, for experimenting. The experiments called for may be such that, humanly speaking, they can no longer be carried out. It could be that the preparation time is prohibitive or that the number of factors that have to be reckoned with in the experiment has become so large that a reasonable amount of insight into the outcome is almost out of the question. Some experiments may be too dangerous or simply impossible. Today, these kinds of experiments can often be simulated with the help of the computer. The computer's calculations take the place of the experiment and generate accurate answers to the questions posed.

Because the computer, in principle, can take over most everything that is determinate in the design process, the emphasis comes to lie more and more on what is typically human in design: freedom in the context of responsibility, the normative dimension of technology, and its translation into the creative or productive imagination. We speak then of computer aided design (CAD).

18. An Overview of the Technological Sciences

When we look at the preceding paragraphs of this chapter, the interrelation and the difference between natural science and technological science have been clearly addressed. Both sciences display analogies in their development, but the intent of each moves them in opposite directions. The natural sciences are directed toward knowledge of physical reality in a universal sense, while the technological sciences aim at concrete, individual, technological realization. Natural scientific claims must, with an eye to attaining universality, be falsifiable. The innovations and discoveries of the technological sciences have to be tested in practice, using specific designs, to see if they are actually going to be fruitful for technological practice.

That is not to say that there are no differences between the various technological sciences. Technological science does form the bridge between the natural sciences and technology (§10). Some technological sciences stand closer to the natural sciences. These we have called the foundational technological sciences. Others focus directly on more tangible matters like materials technology, product technology, and production technology, sometimes also referred to as the applied technological sciences (Teichmann, in Rapp, 1974). Figure 2 provides a partial listing of the basic sciences and many of the technological sciences. It indicates the relation between these basic sciences and the technological sciences as well as the degree to which some of the technological sciences are tied to concrete technology.

The sciences that form the basis of technology are the most general. Then come the foundational technological sciences. Building on these are the sciences dealing with materials, natural resources, and semimanufactures. Finally come the sciences having to do with concrete technological products or processes, the applied technological sciences. Figure 2 does not include philosophy and the social sciences. Given their place at technological universities, they could be considered as contextual disciplines for technology and the technological disciplines.

Applied technological sciences:	
knowledge of technological products	
and production technology:	
- machine tools	- steam turbines
- gas turbines	- coke furnaces
- nuclear reactors	- generators
- automobiles	- aircraft
- radio and television	- telecommunication
- presses	- computers
- bridges	- ships
- agricultural machinery	
knowledge of technological materials (including semimanufactures) or *materials technology:*	
- iron and steel	- aluminum
- oil	- uranium
- gas	- rubber
- paper	- ceramics
- glass	- plastics
- concrete	- textiles
- foodstuffs	- chemicals
Foundational technological sciences:	
- engineering thermodynamics	
- mechanics	- surveying
- materials science	- statics
- calculus	- metallurgy
- fluid mechanics	- machine mechanics
- electronics	- circuit analysis
- weights and measurement	- kinematics
Basic sciences:	
- physics - biophysics	- information science
- biology	- mathematics
- chemistry - biochemistry	- ecology

Figure 2 Overview of the technological sciences
(adapted from Rogers, 1983; the lists are not all-inclusive)

19. Biotechnology

The development of the technology discussed up to this point has moved from materials via energy to information technology. More and more this technological development has become involved with, and supplied the basis for, the development of biotechnology. Tools are no longer the means used in biotechnical formation. In biotechnology, life forms are active in such a way, possibly via manipulation, that they change according to the desires of the technician and subsequently deliver the desired products.

Biotechnology uses microbes, tissue cultures taken originally from plants or animals, and manipulated micro-organisms, plants and animals. It is a mistake to think that biotechnology is of recent vintage. That is simply not true. People have used the biotic processes of micro-organisms for a long time already. Think of baker's yeast, yogurt, cheese, and wine. How these products first arose is often unclear. Through centuries of experience, people have made this technology their own.

About a hundred years ago, the systematic, scientific study of biotic processes was initiated. The results of that investigation, gleaned from life sciences like molecular biology and genetics, lead to the incorporation of biotechnology in fields as diverse as industry (especially in the processing of food), agriculture (particularly bioindustry), health care, environmental protection, and energy production.

The development of biotechnology in the last decades has moved very quickly. New possibilities are the order of the day. One of the most astounding discoveries in biology has been the possibility to play around with heredity. Little pieces of hereditary material, called genes, found in the carrier of hereditary information, the DNA molecule, can be transferred from one organism to another such that cells in the second organism form having new characteristics. With the help of enzymes, pieces of one DNA molecule can be spliced and then recombined in another DNA molecule. Spliced genes, either individually or in groupings, can be inserted in other cells without those cells dying. In this way artificially new cells are made that in turn are able to reproduce themselves. This process is called genetic engineering or recombinant DNA technology and is one application of the same scientific method we discussed previously (§12). The point can then be made that recombinant DNA technology is a counterpart to biotechnology just as modern engineering is a counterpart to technological science.

In 1974, at a conference on molecular biology in Asilomar, California, Paul Berg and Stanley Cohen announced that they were conducting experiments to transfer genes from one organism to another. They related at the same time that they had no idea as to the possible ramifications of their experiments. Scientists themselves became deeply worried about the potential catastrophic consequences, such as the spread of dangerous viruses and bacteria. At that same conference they decided to curtail all research, worldwide, and only under very special conditions to allow further experimentation. From that time on experiments were performed in isolation under very strictly controlled physical and biological conditions. Physical barriers consisted of air locks, filters, and even constructing a research facility on an island. Biological containment required that micro-organisms be "crippled" with the help of recombinant DNA technology such that should these microbes escape the laboratory environment they would disintegrate and destroy themselves completely.

This moratorium on DNA research, in other words, arose within the scientific community itself. Never before had that happened. These circumstances led to a discussion between science and the populace such that society became informed about this new area of research.

As research continued it gradually became apparent that the risks and dangers were not as great as were originally thought. The reins were loosened a bit. Throughout the world today there are set guidelines for quarantined experimentation. In addition, precautionary measures are in place to see to it that genetically manipulated organisms are not introduced outside the laboratory without permission.

The development in biotechnology brings with it unexplored vistas. It is as though people can command and program nature, manipulating it genetically to do their bidding. Crops can be made to resist drought or heat, milk production can be increased, using recombined bacteria in bio-reactors, vaccines can be developed against a sickness like foot-and-mouth disease, a polluted environment can be cleaned up using microbes that are recombined in such a way that they eat up the pollution. New medicines can be made. For example, by inserting the human insulin gene—the gene responsible for the production of insulin—in harmless bacteria found in the human intestine *(Escherichia coli)* these recombined bacteria will produce insulin. This is a great improvement over the only other commercial source of insulin, namely, the pancreas of either cattle or pigs. The same can be said for the production of interferon that is needed for fighting viral infections.

This all sounds rather fantastic and in large part it is. But it is becoming increasingly clear that next to the many advantages, there also can be significant disadvantages to manipulating plant and animal life. Genetic engineering can result in an enormous increase in crop yields. But these crops cannot reproduce themselves, so that new hybrid seed will always be required. The farmer, and here one should think especially of those in the third world, becomes dependent for a large part on the industrial powers. Moreover, the resulting surplus can bring with it considerable economic problems.

Success can also be had in the field of animal breeding and livestock production. But here too there are accompanying ethical problems. How far is too far when it comes to genetic manipulation? Where do the limits lie? May we do with animals simply what we find useful for ourselves? Could it be that animals produced by this technology would suffer under the consequences of manipulation? For example, because of obesity or a disproportionate or weak skeleton? Furthermore, is it possible that by transplanting or cloning embryos we are introducing a technology that depreciates the very nature of these animals?

The ethical questions will obviously be even greater if genetic manipulation would expand to include humans. On the one hand, we may expect that within the context of a responsible medical ethics, medicine will be able to advance, as is already the case with the genetic manipulation of certain organs. On the other hand, human beings can also become the victims of this technology.

With this new technology it is possible via prenatal or even preimplantation diagnosis to determine beforehand any hereditary deformities. This clearly raises many ethical and social problems. What are the implications of this kind of research? Would knowledge of this kind of diagnosis not put insurance companies in a position to exert inappropriate pressure? The problem would be multiplied with a complete catalog of a person's genome. This would include the total complement of genes in an individual, even though the particular significance of each of these genes might not be known. The function of many genes is already known, but for a good number, that is still not the case. That is the agenda of continuing research. The increased ability to predict more and more diseases that will result, will in turn have societal implications for the relations between employee and employers and between client and insurance companies.

In short, given the new possibilities that biotechnology brings, particularly in genetic engineering, we have arrived at a stage in the development of technology that is thoroughly fascinating, but that can also become extremely hazardous. Technology began developing when human beings mastered dead material. But scientific developments have now made human beings themselves the object of technology. A few of the many problems and ethical questions connected with this last phase will be addressed in the fourth chapter.

Chapter Three

PHILOSOPHICAL AND ETHICAL FOUNDATIONS

20. Ethics and Technological Science

In our culture, science and modern technology influence all kinds of areas. Compared to the past, materially speaking, our culture is rich. Our age has seen an unparalleled increase in human power and freedom. At the same time, it cannot be denied that its problems are large and on the rise. For example, because of the predominance of scientific-technological power, agriculture has been disconnected from its ecological and biotic context and the strain between agriculture and nature is intensifying. A rich diversity of flora and fauna are being wiped out, damaging the environment. The pollution of earth, air, and water seriously infringes on the healthful fundamentals of the biosphere. The power of modern technology and the ensuing dynamics of history prove to have a flip side in the threat to nature and man by degrading the environment and destroying our health. The big question that will not go away is whether or not the power of technology is going to get out of hand. Or as Bruno Guardini posed the question: how is our power over the power of technology going to be maintained or reinstated?

In a variety of contexts people are talking today about a cultural crisis. This is due particularly to the fact that our culture has come under the grip of an *overly extended* scientific-technological control. The word "crisis" suggests the fact that we are confronted by an accumulation of problems upon whose solutions few can agree.

There has been a good deal of reflection from a variety of philosophical positions on ethical issues related to technological science. In our survey of philosophical perspectives (§§8 and 9), for example, we saw the extent to which cultural philosophical discussion has focused not only on philosophical questions concerning technological science, but on ethical problems as well. In this chapter I will want to contribute to that philosophical, ethical reflection. In doing so I will be discussing an approach that is more radical and more integral than is usually proposed.

In that context, I will analyze the influence and pivotal position of technological science. This is necessary because previous discussions have shown that within technology, tension exists between science, on the one hand, which by nature is always universal, constant, abstract, and logically coherent, and reality, on the other, which is singular, continually changing, replete, and everything but rational. Given the extensive influence of the scientific-technological control of culture, we cannot exclude the possibility that causes for the crisis can be found in this tension.

Problems indicated in Chapter I (§5) could very well be related to an undisclosed prescientific or extra-scientific dynamic and view of science and to its structural consequences in the repercussions of technology's scientific control and its influence in so many areas of culture, such as agriculture, economics, health care, and politics. Through scientific-technological control, science is having more and more influence on society.

Moreover, it certainly seems that there is a blind spot when it comes to considering these preliminary questions. The silence must be broken regarding the philosophical and ethical foundations having to do with what most deeply moves human beings.

I want to emphasize that the philosophical and ethical approach I am taking proceeds from the perspective of reformational philosophy

(Dooyeweerd, 1953; Kalsbeek, 1975). In addition to its analyses of the structure of reality, particular attention has been given in this philosophical tradition to the deeper driving forces, or motives, in the cultural activity of human beings. After discussing some of these we will also want to look at whether reformational philosophy's theory of the nature of things contains a normative context for scientific-technological control which might suggest a way to understand the problems of technological culture.

Before proceeding I want to clarify what I understand ethics to be. Ethics can be defined as the science or discipline that deals with what is good or responsible human activity. It also involves making explicit the body of norms and values that make an activity "good" or "responsible," as well as the motivation that lies at the basis of that activity. In this chapter I will speak of ethics in this last sense. I will be talking about setting an ethically good, and hence responsible, meaningful direction for developing culture and, in so doing, for developing both technological science and modern technology in its relation to the various sectors of society. We have to seek out a normative context within which modern technology and technological science, to which modern technology is attuned, can develop responsibly; a normative context that opposes the overestimation, or absolutization, and blanket imposition of the scientific design method.

21. Technicism, Utilitarianism, and Instrumentalism

Human power has increased tremendously, but human beings do not seem able to handle this power. Using science and technology, we have conjured up very serious problems which now seem beyond our grasp. We stand powerless over against the deterioration of our inner cities, the devastating character of nuclear war, the problem of aid to third world countries, the degradation of nature and the pollution of the environment, the many problems tied to unemployment, the many irrational reactions to these demoralizing problems, for example, in the use of drugs, the growing loneliness of people and their increasing sense of alienation, and the list goes on.

In our quest for the causes of the present cultural crisis, we are going to begin with a discussion of the spiritual and historical roots of the development of science and of the technology to which it has given rise.

Following more or less the path of a growing number of philosophers, like Heidegger, Ellul, Sachsse, Staudinger, Horkheimer, and Ihde, many have come to recognize *technicism* as the *predominate* spiritual force in the development of science and technology. It is important to note, however, that this driving force, while predominant, has not been the only one. The fact that there have been others makes it possible for us to indicate another, more acceptable route for the future.

Technicism is the human pretension to be able to autonomously bend all of reality to our will through the use of scientific-technological control and in doing so to solve any problem that arises and hence to guarantee material progress. The assurance is that even the problems technicism creates can in turn be mastered with new or proven scientific-technological solutions. Spengler had this conviction in mind when he spoke about a "faith in technology" as the stimulus for technological development.

To avoid any misunderstanding: my intent in discussing "technicism" is certainly not to dismiss technological science and modern technology. My critique of technicism is aimed at rejecting the *predominance* of technological science, while promoting its ability to be of *service*; to dissuade people from *overestimating* technological power, while encouraging *responsible, meaningful* technology.

Technicism has come to dominate more and more the ethos, the guiding principles of human knowledge and activity, in science and technology. In talking about technicism, then, we are really talking about a religious conviction that manifests itself in what many have referred to as a "faith in progress." That confidence, as a faith in material advancement, is the driving force behind the expansive and absolutized growth of science and technology. As far as this faith is concerned, all of reality is just "stuff" awaiting the formative and creative power of humankind. Under the influence of technicism, the successful scientific design method of modern technology (see §11) is converted into an absolute. Furthermore, there seems to be something imperialistic about this method as it also begins to make its presence felt in areas of culture not directly related to technology. Considered an absolute, it proceeds to impose its irrefutable mark on everything under its sway, instead of rendering a service within the various sectors of society.

The influence of technicism is also readily apparent in the misconceived notion that one need not distinguish between natural science, technological science, and technology. This misrepresentation quietly harbors the view that science is the *instrument* for technological control. In this instrumentalistic conception of natural and technological science is hidden an *utilitarian* ethos. What can be made, must be made, with an eye to enhancing the utility of what is made, for as many people as is possible.

Let us take a closer look at each of these three.

a. The influence of technicism comes to expression in Western culture for the first time in the spiritual movement of the Renaissance. The sweep of technicism broadens from the moment that the Renaissance put its stamp on Western philosophy and the development of science. Later, due to the influence of the Enlightenment, of positivism, and of pragmatism, technicism became dominant in our culture.

The influence of technicism has not always been so great. Through the influence of Christian movements like the Reformation, the Counter-reformation, and the Great Awakening, but also owing to the influence of Romanticism and, more recently, of the Counter-culture movement, the promulgation and application of technicism has met with some resistance. As for the question whether Christendom is culpable for technicism and its consequences, much has been written. Historically, the influence of the Reformation on the development of the natural sciences is unmistakable. But yet it was philosophy, which became predominantly technicistic under the influence of the Renaissance and later humanism, that more and more put a stamp on the development of science and technology.

The father of this technicism is in a certain sense René Descartes. He claimed that the rules and laws of mechanics are the same as the rules and laws of nature. Basic to his philosophy of nature is the paradigm of the automaton, that is, of the machine. "Nature is a machine, as easy to understand as clocks and automata, when one but investigates it carefully enough," says Descartes. What this eventually implies is that nature can be figured out and controlled. After all, man is "the master and owner of nature." That is why Descartes, already, considered plants and animals as things to be manipulated.

Under the leadership of later philosophers, this technicism became stronger and its influence began to spread into society and politics. Hobbes raised ideas about a technocratic state and Comte wanted to develop techniques to control societal processes. In our day, with genetic engineering, we realize that humankind itself can become the object of this technicism (see §28).

b. Technicism expresses an ethical conviction. This means that under the influence of technicism the development of science and technology has been set on a specific course. This direction corresponds to what in ethics is called *utilitarianism*. In utilitarianism those goals of science and technology are considered good which are useful to as many people as possible. For example, in the case of scientific-technological control, people pursue those goals that comport well with technicism. Utilitarianism has an eye only for material prosperity, long life, happiness, and the like. Again to avoid any misunderstanding: these goals in themselves are certainly valuable. The ethical problem concealed in technicism and its bedfellow utilitarianism is that these aspirations are converted into absolutes and that these absolutized goals have become definitive for scientific and technological activity. That is to say, the manner in which and the means by which these ends are achieved through scientific-technological control are not given any, or at least not enough, attention. At the same time, it is the ends that sanction the means to scientific-technological control.

Utilitarianism's narrow vision of a "good" and "responsible" technological science, and the technology that comes with it, has led to many of the problems we have mentioned (see §3). The ethical attitude of utilitarianism results in a one-sided technology that is harmful to people, nature, the environment, and society. By means of scientific-technological control, nature is reduced to the use humans can make of it. What we have to do to use it as we wish and how we use it are of little concern. This course proves to be destructive. Anything that does not fit the parameter of technicism is misjudged or disregarded. Nature no longer seems to have a value or meaning of its own and human beings are too often reduced to machine components or to a complex of consumers.

The outcome of an unlimited scientific-technological control of inanimate and animate nature can lead to the dehumanizing of labor, the atomizing of society, and the degradation of nature. The biosphere in which we live forms a highly complex, singular whole. When this is not taken into account the result will be the gradual reduction of plant and animal species.

But destruction is not the only consequence. Due to limitless industrialization, to modern scientific-technological practices in agriculture, and, not to be forgotten, due to modern means of transportation, earth, water, and air pollution are increasing at an alarming rate. And as nature becomes less stable, the desire for technological control grows, generating a dangerous and vicious circle.

c. To better understand the consequences of technicism and utilitarianism we must examine the characteristics of natural science and the *scientific design method* (see §11). Technicism and utilitarianism project, via the *instrumentalistic* maneuvering of science and method, their characteristics onto the breadth of reality, such that it is remodeled, reduced, and disfigured.

Technicism, we saw, is the attempt to create a world that is molded to the human hand. When science and the scientific method are absolutized, we can refer to this method as one of *destruction and reconstruction*. Without taking into account the given order of reality, its structural diversity, and the mutual interrelationships within that reality, everything, including their smallest components, is broken down or taken apart with strict consistency and then, using these elements, put back together again as we see fit. Converting the method of systems theory into an absolute leads to the same result (see §11). This philosophic and scientific "creative" renovation of reality gives rise to a mechanized world picture (Dijksterhuis, 1986). Later this inventive reconstruction results in the practise of scientific-technological control and a corresponding technological world picture.

It is only during and after the industrial revolution that technicism comes to cultural fruition. With the rise of industrialization all of reality is placed within the grasp of an overextended, scientific-technological control.

The substance and consequences of technicism in the guise of instrumentalism can best be approached from the spiritual attitude of the Enlightenment which arose in the eighteen century. In this movement, the spirit of the Renaissance, to wit, an all-out trust in the human ability to renew life, is connected with the development of the natural sciences. The arrogance of human autonomy, man as Prometheus, someone who needs nothing more than himself, attaches itself to science. Inspired by the thriving development of natural science, the superhuman of the Enlightenment thinks he can solve all problems and with the help of the natural sciences, and later technological science, bring new life to society.

Owing to the influence of the Enlightenment, and via that of positivism and pragmatism, the ethos of many scientists today is characterized by technicism and utilitarianism. The essence of science is interpreted technicistically. Science becomes an instrument of control and is used to resolve any presently pressing problems. Utilitarianism affirms and reinforces this process with the promise of unlimited material prosperity. Because people no longer see in this process the limits and limitations of science, the way is open for a limitless scientific-technological manipulation of reality (Staudinger, 1976).

The preponderant role of scientific utilitarian thinking means that values and norms that conflict with people's wants and desires are simply dismissed. That is why reality's pregiven meaning and coherence are disregarded more and more in science.

Within that spiritual climate it was easy for positivism and pragmatism to stamp out any resistance to instrumentalism, the unbounded scientific-technological control of reality. Everything there is, is considered as material to be controlled. The worldview that fits with this assumption we call materialism.

Consequences of Technicism, Utilitarianism, and Instrumentalism

Initially the results of this development are impressive. Actually, they fascinate many still today and are often perceived as a means to unprecedented progress and material prosperity (Von Weizsäcker, 1964). Yet, it is important to realize that from the very beginning of this development there has been talk of a latent lack of direction. Only later did some begin to listen and take note.

On the one hand, human beings are assumed to be the only authority to which one can appeal, and as such, the supreme beings within reality. But on the other hand, seen from the side of science, human beings are the result of a material development that has been determined by science. Man is "lord and master." But as the product of evolutionary coincidence, he is also victim. Because human beings do not want to draw this conclusion of being victims of chance, the quality of mastery remains dominant. Nevertheless, in the face of looming problems, it cannot be denied that pessimism and uncertainty are present and growing as things continue to develop. The lack of direction mentioned above is becoming manifest.

That was hardly the case in the eighteenth century for a materialist like Baron d'Holbach, to mention just one example. He says that the world shows itself to be nothing more than matter and motion, connected in an infinite chain of causes and effects, and that our scientific knowledge of the same provides humankind the key it needs to be able to control reality. In the early nineteenth century, the mathematician Pierre-Simon Laplace was convinced that we would be able to get both reality and the future under control by means of science. Many representatives of the Enlightenment, including Immanuel Kant, thought the same way. Although he was not a materialist, he was still strongly influenced by technicism. With an eye to controlling history, he called for a Kepler or a Newton who would discover the natural laws of history. The ultimate goal that Kant, as well as d'Holbach, had in mind was a congruence of physics and morality. This would imply that by means of instrumentalization natural science would come to function as the standard for practical activity.

When this process is affirmed culturally and begins to spread, and when, as a result, problems and threats, like the degradation of nature and environmental pollution, increase, and when people begin to realize that they cannot simply appeal to science and to scientific-technological control for a solution, then, as they come to see their predicament as the result of their own handiwork, they will be left without a clue as to what to do. Today, our culture's lack of direction manifests itself in the confusion that reigns among suggestions for a way out of the crisis in which we presently find ourselves.

We have to recognize that under the influence of technicism the world around us has been forced increasingly to conform to the characteristics of science. The instrumental use of science that technicism encourages, to wit, *instrumentalism*, has tried its best to fine-tune reality according to the sounds of science.

To understand the effects of unlimited, presumptuous scientific-technological control we must examine the structure of science.

One hallmark of scientific knowledge is that it is abstract. When scientific knowledge, due to the influence of the Enlightenment, is elevated to an instrument of control and people no longer have an eye for the reductions implicit to the abstractions of science, these abstractions, through the irresponsible use of science, begin to mark a culture. That is to say, because of their inherent reductions, the wholesale use and resolute instrumental employment of the abstractions of science can lead to the degradation of reality and, by mutilating its meaning, even ultimately destroy it.

Which *abstractions* characterize science? To come to scientific knowledge one must follow the path of *analysis and abstraction*. Along that path (which is what "method" means literally) more is going on than analyzing and abstracting. For example, we form hypotheses and we experiment. But analyzing and abstracting are certainly the most prominent activities. The scientist distinguishes within a many-faceted reality among a great diversity of functions or aspects, for example, the physical, biotic, and economic aspects. To investigate this diversity scientifically, the scientist first abstracts from this coherence of aspects just one aspect or function. The second abstraction is that the scientist, within the field of the initial, functional abstraction, brackets out the particular and attends to what is general or universal in that field. The third abstraction is that scientists distance themselves from sensible, perceptible reality and turn to the *laws* that *hold* for this reality. That is why science is often, and correctly, compared to a "map." One arrives, for example, at the formulation of a natural law with which the researched phenomena comply. A fourth and final abstraction is that the scientist is supposed to ignore any benefit to or interests of oneself or others. Which is to say that "science must be disinterested" (Van Melsen, 1970). Failing to honor this abstraction as the standard or norm for scientific research is to give tacit consent to utilitarianism and implicit approval to having science simply benefit human interest through scientific-technological control.

Within the context of these abstractions, the scientist is seeking *to gain theoretic knowledge.*

Given the above, we can say that science follows the route of these four abstractions. To arrive at good scientific results, every science must learn to follow this route. Exaggerated slightly, this means that the scientist proceeds wearing blinders. This is inherent to science. The scientist ought to be conscious of this fact. He brackets out a great deal to which he gives no further attention and concentrates on what remains within those brackets. The scientific knowledge gained along the route of analysis and abstraction is then incorporated—using logic—into a *system* of scientific knowledge. That is also why we say that scientific knowledge is *logically coherent* knowledge. When we transpose this scientific knowledge into mathematical relations (what we could call a second level abstraction), we can manipulate it by using information technology, which in turn positions us to manipulate the reality that one originally set out to study.

To get a good view of the features and characteristics of scientific knowledge, we need to set these four abstractions in a row. Scientific knowledge is universal, functional, and, as knowledge of the law for whatever we are studying, valid. These properties, as Van Riessen (1970) has rightly noted, are *incongruent* with the fundamental characteristics of the full reality we experience. In *that* reality, everything is different, interrelated with everything else, and ever-changing.

When we do not recognize that science distances itself by means of abstractions from the fullness of experienced reality, as happens under the influence of the Enlightenment, then a good deal of the fullness of that reality will be lost with the instrumental use of scientific knowledge.

A simple example can make this clear. In theory we know how to distribute four apples among four children. But in reality doing so is not always so easy. Not all apples are created equal and the preferences of children are not only different, but change continually. Our calculation did not take the size, color, or condition of the apples into account. But in reality, it is usually these factors that determine the choice of children. Obviously, applying the division of four by four has no earth-shaking repercussions. But it is different when through the "instrumentalization" of science, for example, in agriculture, industry, or health care, the fullness of reality is converted into the abstract, qualified context of science. Influenced by technicism, people have claimed that science can fully grasp reality. They are convinced that science can solve any problem that arises. In other words, science's "map" is equated with reality. Aided by science, all of reality is molded to the human will with an eye to increasing human power and generating unprecedented and unlimited prosperity. Science no longer serves everyday life, but is beginning to influence and even to dominate daily life. This especially happens when the scientific design method becomes a will-less pawn of economic and political powers. These powers exploit the scientific-technological power of control to bolster themselves and the materialism present in a culture.

But has modern systems analysis not seen and sought to remedy many of these dangers? In one sense this is true. But giving attention to the whole as a system does not take away from the fact that it is still a constructed system, a *simulated* whole, and, as such, abstract. Systems analysis does resist the world picture nurtured by technicism. But however important and valuable its revisions have been, systems thinking itself remains tied to a technological world picture (Ellul, 1980). Invoking systems analysis, as the new way of thinking to resolve the cultural problems of our day, may possibly bring temporary relief. But remembering the structure of systems thinking and the limitations inherent to it (see §§8 and 12), this uncritical appeal will, in the long run, confront us in a more intense manner with both old and new problems. Ultimately, we model reality according to the standard of the system we have constructed. As we nullify earlier reductions, we see new ones. An apparent solution ends up producing a convoluted complexity without prospect. Furthermore, systems thinking does not consistently halt growing materialism or faith in progress. To the extent that the theme of progress is replaced by that of survival, the cultural dynamic can indeed be slowed down. But, even then, the question is whether this deceleration is possible given the trends of mass culture or whether it will be acceptable to those who have yet to enjoy the fruits of modern technology and scientific-technological control. Therefore, for the moment it seems that, all things considered, a systems analysis approach advances a technological culture more than it slows it down or changes its direction.

22. Making Everything Technological: Technologizing

We noted above that science and technology are taken up into an interrelated complex in which economics and politics also play an important part. Technicism is the driving force in this complex. Bolstered by utilitarianism, technicism, as we saw, has turned science into an instrument of control. In doing so the possibilities of technology are overestimated, so that now, influenced by the imperative of technological perfection, everything that can be made is made. The eventual result is that everything becomes technologized.

Technicism, which wants to guarantee material happiness and prosperity, also leaves its mark on economics and politics. Economics becomes the economics of material progress, in which the power of money and material gain dominate (Goudzwaard, 1980). Materialistic politics enhances this process and in so doing, plays into technicism. Politics, as many have noted, is becoming increasingly technocratic, meaning that the method of scientific-technological control also influences politics.

People outside these cultural sectors, given an insatiable desire for material prosperity, eagerly buy into this development. They are convinced that they can only gain by it. Other sectors of society, like agriculture, health care, and a diversity of organizations, have to learn to put up with the influence of a clustering of cultural powers. Although the opposite was promised, here too the convergence of powers proves a disruptive influence and accelerates the crisis within culture. For example, farmers tend to overproduce, on the one hand, and to impoverish the soil, on the other.

When this development began on a small scale, the ramifications were hardly serious. In fact, short term effects promised nothing but success. That is also what makes it so appealing. Building a kind of "alter-creation" in which human beings would be lord and master over everything and be certain of untold material wealth, blinds one to what is actually going on here.

As the process of technologizing increases in extent and magnitude, something that reflects the preponderance of science, negative effects become apparent and threaten to predominate. Reality is shaped according to a reduced, logically coherent network. The abstract framework becomes so dominant that the fullness of reality is broken into atomized and functional units or reduced to artificially constructed systems. The absolutizing of the artificial, from flowers to intelligence, makes reality more matter of fact, maims its meaning character, and rips reality into component parts. The problems of a technological paradise are great.

This process of making everything technological is evident most everywhere: in the dehumanization of labor, in the devastation of nature, in the pollution of the environment, in problems related to nuclear energy, in making society information-compatible, in problems related to farming and meat production, and in problems surrounding the genetic engineering of humans. In a certain sense, the magnitude of these problems increases as we move down this list. Fortunately, we have not yet been confronted as such with the full potential of many of these problems; in part because the practical expansion of biotechnology and genetic engineering techniques is still subject to clear limitations. However, we must weigh more fully the eventual consequences of their unrestricted application.

Two examples can help acquaint us with the significance and influence of technicism and its consequences in the technologizing process. In order to also make clear that in rejecting technicism and the attempt to technologize everything we are not dismissing the value of technological science and technology, we will examine briefly a responsible development. The first example has to do with the development of nuclear energy; the second with the information age.

a. Within a context defined by technicism nothing is allowed to hamper the development of nuclear energy. In contrast, the large scale of the technology used to generate energy seems to be the source of so many advantages and to guarantee increasing prosperity. It is this materialism that initially sees to it that nuclear energy must be had as cheaply as possible.

Since Three Mile Island and Chernobyl, it is obvious that grim catastrophes can happen with atomic energy and that their repercussions are immeasurable. Scientific-technological control is not nearly as safe as many have claimed. If the control mechanisms should fail at a certain point, the aftermath could be devastating for a long time and for a large area. When we also take into account the unresolved problem of storing radioactive nuclear waste, it is clear that we have yet to arrive at a responsible final phase as far as this technology is concerned.

The waste products of nuclear reactors pose significant health problems for people today as well as for those in the (distant) future. That is why this radioactive refuse must be able to be stored securely for a long period of time, often for a good number of centuries. Does this responsibility not go beyond what man can handle? And even more primary is the question whether producing these hazardous wastes is permissible. Every cultural activity comes with a price. But can the phenomenon of radioactive waste as such, to which so many objections and concerns are tied, be reconciled with a correctly *normed*, ethically acceptable, technological development? In other words, can we speak of a good and proper technology if its long term products are hazardous? If they remain hazardous for so long that it is simply impossible for the designers of this technology, the technological scientists, to bear responsibility for them?

This peril in the development of nuclear energy is only accentuated when we consider that the development of atomic energy goes hand in hand with possibly producing nuclear arms. This, by itself, can have significant detrimental consequences. In addition, we have to figure that, politically, strong protective measures will have to be taken. The political realm, confronted with the drawbacks of technicism and the desire to make everything technological, is becoming more and more technocratic and, as a result, is often at odds with democracy. This development has elicited a number of strong reactions. For example, virtually every new nuclear power plant brought on line or even only contemplated in the last fifteen to twenty years has met with demonstrations of protest.

In order to constrain a growing cultural adversity to large scale technology, energy policy makers must find an alternate route. Instead of technologizing everything, they will have to consider conserving energy and developing alternative energy sources, like solar and geothermal energy or energy derived from wind and water power or biomass. These forms of energy make a culture less dependent on one particular source and hence more stable. With a more varied development of energy, fewer risks in space and time are taken than is the case with the unhindered development of nuclear energy.

Only after these initiatives have been realized could it perhaps be possible to talk about a limited and stringently safeguarded development of atomic energy. The suggestion of C. F. von Weizsäcker, advisor for the German government regarding energy policy, to build maximum security nuclear reactors underground deserves further attention.

b. Many of the social sciences seek to emulate the natural sciences in method and outcome. When technicism's procedure of brutish deconstruction and headstrong reconstruction of given reality extends by way of the natural science to the social sciences and then to *society*, the disastrous consequences of that method will also begin to affect society itself. This disruption will manifest itself in fragmented communities and isolated individuals who have been alienated from their original place in society. Was that not often, even still today, the reason for the dehumanizing of labor in modern factories? Is not the same thing happening to many today who work the land? Many farmers are being robbed of their freedom and responsibility. Some are being forced into early retirement. Others are coerced to change careers. Many feel they have been alienated from their land and animals and from nature.

Modern society is entering the information age. Interactive television, computerized education, and electronic banking, news reporting, and mail order shopping, all supported by electronic information systems, will soon be considered commonplace. It is all just one more step in the process of technologizing society. However, the hope that the newest technology is ushering in a new future stands in the way of any critical distance with respect to information technology. Given their conviction that if it can be made it should be made, combined with a faith in progress, many uncritically assume that information technology will be the panacea for all of our past and future problems, facilitating a perfect, technological streamlining of society. But the other side of the coin will be human alienation, a feeling of disintegration, and a lack of direction.

The desire to make everything technological is evident in education, politics, futurology, and in the modern workplace (see §26). What follows will focus on the workplace.

Human labor is multifaceted. Because of the influence of technicism, work has become more and more related to business and technology. The production sector, and the services that support it, tend to be considered the only kind of valuable work. That is to say, imaginative fields, work in the social service sector, and works of compassion are depreciated. Vocations that are devoted to the transmission of culture, of a culture's norms and values, also receive less attention. As a result, secondary education is experiencing a shift toward fields related to business and technology, while other educational fields are left in the lurch. For culture this means a significant loss.

The unrestrained effort to technologize work will eventually pose a significant problem, certainly in the production sector. In principle, this kind of work can for the most part be automated. When this happens many will become unemployed. Unfortunately, most of these people will not understand that other valuable kinds of work exist. C. J. Dippel quite rightly speaks in this connection about the missed chances of modern technology. His point is that technology offers so many new possibilities for all kinds of work, but that a narrow view of technological development is yielding just the opposite.

Although it should be clear from the above, I want to emphasize again that protesting against the move to make everything technological does not imply an objection to technology as such. The point is rather to defend a responsible technological development that does justice to its meaning and relative importance. People are going to have to recognize more and more that technological science must serve humanity. Its place is to support and not to suppress (see §§24 and 25).

In summary: the dubious process of making everything technological, of which we have given two telling examples, is based on absolutizing scientific-technological possibilities, allowing them to rule life.

Influenced by the materialism of producer and consumer, the move to technologize has been pushed by economic and political forces that themselves put their stock in technological power. In agriculture, for example, technology contributes to overproduction. To keep ahead of progress, the renewal of scientific-technological control finds itself in an upwards spiral that both is world-wide and, given the effects of technologizing, can become catastrophic. Global contrasts, as well, become the more biting. While forty percent of the world's population lives in poverty and suffers from malnutrition, people in the United States, for example, spend fifteen billion dollars per year on losing weight, and underscoring the sharp contrast, an extra twenty-two billion on beauty aids.

In the meantime, economic and political powers, aided by biotechnology and the newest information technology, are allowed to increase in scale and intensity. As they do so, one's ability to scrutinize them decreases. Together with the tangible problems of technologizing, they are defining the crisis within culture and setting a course that knows no direction.

23. Science and Wisdom

Up to this point we have seen that technicism influenced science to became an instrument of control. This instrumentalism has been very successful in bringing about material prosperity. But at the same time, its large scale application, absolutization, and imperialism have led to the problems of technologizing.

The science that plays such an important role in modern technology need not be renounced. Modern technology in its entirety would be impossible without that science. What we need is a different view of science, one in which science is not allowed to dominate, but is nurtured with an eye to service. The place of science should be to assist and bolster. We have to let go of the mistaken notion that with science at our side we, as "lord and master" (Descartes), can mold reality as we see fit. In contrast, I would maintain that as stewards of creation we are obliged to insist that science be of service in its dealings with reality and, likewise, in technology.

Within that context science can contribute to a growth in wisdom. That is to say, we have to recognize that scientific knowledge is carried by experiential knowledge, which is full and concrete. Science ought once again to stand in a correct relation to reality in its fullness. We should integrate it into the fullness of the reality we experience everyday. Doing so will deepen that experience. In this way science contributes to an increasingly extensive understanding. That is what *growing in wisdom* is all about. Science will no longer function as the main thoroughfare in cultural activity, or even when it comes to technology, but as a helpful service road. The difference this would make in technology is that it would not only be a matter of building, and then building some more, following the technological mandate that what can be made must be made, but it would be just as much a matter of preserving and protecting nature. Wisdom, in the sense of a growing, comprehensive insight, will also lead to creative and circumspect activity in technology.

In order for us to grow in wisdom and for science to remain of service, we have to acknowledge that we must integrate the abstractions discussed previously into prescientific, or extra-scientific, experience, so that the limitations of scientific knowing can be lifted. When we integrate the scientific knowledge we bracketed out along the scientific path, we must once again take it into account. That holds first of all for the abstraction of a functional field with respect to the interrelatedness of reality. We must do justice to all of reality and it exhibits more aspects than the aspect investigated by one particular science. The approach of any one science is then insufficient for solving some problem, certainly today, when so many complex problems exist. Problems require a *multidisciplinary* approach. And yet multidisciplinary collaboration, as the sum total of the knowledge coming from the various special sciences, is still no guarantee that justice will be done in a concrete situation. We understand this fact only when, in contrasting the generalizing approach of multidisciplinary collaboration with the approach of specialists in their discipline, we realize that scientific knowledge is characterized by more than one abstraction. Within its functional field, a science focuses only on what is universal. This abstraction, and the universal scientific knowledge that ensues, we often mistake for knowledge of concrete reality. This reality consists of only particular, discrete, everyday entities. Restoring the abstraction of what is universal will ultimately contribute to an enriched knowledge of the singular reality of everyday. But the richness of reality may not be relegated to abstracted universality nor may science claim that it can grasp what is unique and may control what it can grasp. Scientific knowledge may not be blinded by its instrumental use, but must seek to serve our knowledge of the particular things around us and what we do with them. When we recognize, for example, the distinctiveness of plants and animals, we will want to set limits on our desire to submit them to our scientific-technological control.

The map that science provides may not be equated with reality. As map, however, science can provide us with a better sense of orientation. This orientation is pre-scientific. The full knowledge of everyday life will be enriched by science and help us become more responsible people.

In this connection we should attend separately to the view that a different scientific approach, for example that of an integrated systems approach, solves the problems we have been addressing. This new approach would have to see to it that both technology's development and the study of societal (cultural, economic, political) consequences are investigated in relation to each other. This new science would have to nurture our knowledge and ability as well as the required socio-practical insight. The advantage of doing so is that we can pose questions concerning norms and ethics clearly.

Many, including someone like the German philosopher Jürgen Habermas, assume that we can discuss normative and ethical questions rationally. But there lies a problem in this hope for communal, rational communication. There is simply no way around the fact that a fuller or broader rationality that includes more sides and more areas, a rationality of social reflection that replaces the present rationality of the scientific-technological dimension, will still remain dependent on rationality. The only answer that can be given as to the basis or grounds for this rationality is, ultimately, rationality itself.

To dispel the pretended self-sufficiency of rationality of whatever kind, reformational philosophy acknowledges that reason is rooted in a given, normative structure of reality that is neither subjective nor arbitrary. This normative structure precedes any discussion regarding rationality and therefore is crucial for determining the place of reason as well as that of science. It also indicates the subordinate place of science and its significance for scientific-technological control and for technology.

It is actually the given, normative structure of reality that points us in a responsible direction and that will help us out of the crisis of the technological culture without having to abandon science, technological science, scientific control, and technology. They will have to conform to the normative limits, which together give a meaningful direction for culture.

24. Driving Forces and an Integral Framework of Norms

What can be said about this given, normative structure? This structure, which the Christian philosopher Johan Mekkes often referred to as the "dynamic of creation," consists of a large number of distinct but interrelated norm principles to which all human beings are obliged to *respond*. The direction of this *guide* for life is "horizontal" (or immanent) as well as "vertical" (or transcendental). Human beings are required to work out these norm principles into norms that can function as signposts for responsible cultural activity.

This *response structure* ought also to help technological science serve the development of modern technology. Usually, that is not the case. Moved by the spirit of technicism and a naive view of progress, what can be made, is made. Few seem to notice what is actually happening. The result is that raw materials are wasted, ecosystems are damaged, jobs are dehumanized, and no account is taken of future generations.

When one recognizes that human beings are meant to do their work in culture *coram Deo*, before the face of God, and be guided in doing so by the normative character of the dynamics of creation and one acknowledges that human beings are not, as in technicism, the center of reality, but must in cultural activities deny themselves, loving God and neighbor, then the intent of their various cultural activities receives a different content. Instead of the central motive of power, in which everything we do revolves around ourselves, we have the central motive of love, which produces divergence in our various cultural activities.

Setting our sights by the normative structure, implies a broad passage and, at the same time, a fixed course. This path begins with accepting the *motief*, mentioned above, that must pervade science and technology, namely, that of *wisdom* and *building and preservation*. "Wisdom" and "preservation" have often been forgotten. Were that not so, biology and ecology would have been accepted as foundational sciences for technological science (see §10) a long time ago. With "preservation" we are talking, for example, about preserving the integrity of nature and maintaining a healthy biosphere. Contemporary developments in technology often lead to the technologizing of nature. All too often there is no sense that technology needs a more comprehensive scientific basis and that this basis, as a contributing factor to a growth in wisdom, leads to creative and judicious activities within modern technology.

Wisdom and the motief of build and preserve dictates that modern technology *ought* to involve itself with the immediate situation in which people find themselves. This includes attending to nature, the environment, and even, for example, the landscape. Modern technology ought to be *adaptable* and *ecologically responsible technology*. Where disruption has occurred, we must make the greatest possible effort to restore things. That in no way means reverting to the technology of the craftsman. Compared to present developments, an adaptable technology ought to be expanded. This differentiation in technological development clearly will also require a cultural dimension. Technology ought not to conflict with the state of cultural development, and the rich variation within it, but to tie into that development in such a way that technology enhances a culture. Unfortunately, the opposite is often true in third world countries. But even in industrial countries there are serious problems. Sometimes technological wastes harm the environment and human beings. We must realize, and that holds in the first place for the engineer as technological scientist and technician, that hazardous wastes cannot be accepted in responsible technology. The engineer's task includes responsibly disposing of technological waste. Sometimes that proves to be very successful. One example: for a long time people in the ice cream and cheese industries had no solution for the obnoxious smelling and sewage contaminating watery by-product called "whey." Later they found that what was first considered waste could be transformed into excellent feed for hogs. This was, of course, both ecologically and economically advantageous.

If technology proceeds from the starting blocks mentioned above, modern technology and alternative technologies, what E. F. Schumacher (1973 and 1978) calls intermediary technologies, should not oppose each other, but can complement each other. Presently these approaches are mutually exclusive. On the one hand, some over-estimate the effects of scientific-technological control and the scientific design method; on the other hand, others, motivated by a romantic view of nature, plead, often as a reaction to present hazards, for alternative technologies. The same tension is evident, for example, between a scientific-technological approach to agriculture and the movement towards organic farming. This tension, caused by scientific-technological control and the problems it raises, elicits reactions. This tension can be avoided and abated when the development of technology and technological control simultaneously satisfy the standards set by many interrelated norm principles.

We will briefly review these norm principles and their articulation in norms and discuss a few key points about their significance for modern technology.

The *cultural-historical norm* is the standard of differentiation and integration, of continuity and discontinuity, of centralization and decentralization, of large scale and small scale, of uniformity and pluriformity. The various components of this norm may not be taken as opposites. Setting the sights of the scientific design method and technology by one side of this norm, will put them on a one-sided and eventually dangerous course. When the focus falls on large scale integration, continuity, centralization, and uniformity, the problems of technologizing will by and large be the result. In addition, the technological development that follows this skewed target will, in turn, result in overproduction and redundancy.

When both components of the cultural-historical norm are met, technology develops more steadily, leaving room for creativity and innovation in new inventions. When we satisfy the cultural-historical norm, technology will be adapted to the existing culture, renovation will become possible, and a stimulatingly diverse technology can arise. For example, rather than expecting nuclear reactors to answer all our energy problems, people will attend to as many different energy resources as possible (see §22). But we must also observe additional norm principles and articulate them responsibly.

Technology unfolds, and its meaning deepens, when we obey the lingual and social norms. The lingual norm is that of *information* or openness. This means that we must clearly inform society about technological renovations. Only then can those working with the technology, or purchasing its products, responsibly evaluate things and make decisions. That is why this norm is closely connected to the social norm, namely, to the norm of *communication* or interaction. Without open communication, it is impossible for those who participate in technology to fulfill their communal and individual responsibilities. Those who develop technology have the responsibility to take the time to inform and communicate with the public. Quite obviously, people in the technological sciences, in particular, have to honor these norms of information and communication for these are the ones who stand, as it were, at the cradle of every new technological development.

The economic norm of *stewardship* must also be honored, but not to the exclusion of the others. As it applies to technology we are talking here about the matter of *efficiency*. Its presently one-sided application is due especially to the prevailing influence of economic theory within industry. Influenced by technicism, industry has focussed far too narrowly on a natural scientific sense of efficiency, in which only those goods that can be expressed in monetary terms determine what has value. We must integrate the economic norm into an integral framework of norms. We must apply it not only to production, but to the use of raw materials, energy, nature, the environment, landscapes, animals, and even of the people involved in developing technology. Problems arise when we limit the economic norm exclusively to techniques of production. Technology and the economics of industry develop into a jumble. But when we apply the economic norm, in conjunction with the other norms, simultaneously across the board, we can prevent a kind of *overdevelopment* in producing surpluses and restore a kind of *underdevelopment* in being stewardly in dealing with nature. We will begin to attend more to nature and the environment, to the scarcity of raw materials and energy, and will come to see that people are much more than their ability to function economically within a technological context. We must acknowledge the responsibility of employer and employee.

The normative development of technology is advanced when we abide by the norm of *harmony*. Non-essential surpluses, the decadence of the materialistic way of life, and the degradation of nature, make it abundantly clear that this norm is not being followed. If it were, given the other norms we have discussed, people would realize that technology ought to be developed in a balanced manner. This norm also requires that we introduce and incorporate new technology in a revolutionary fashion. Doing so can bring with it societal unrest and a loss of communal support. We must also consider this norm of harmony in the multifaceted interrelationships between nature, people, culture, and technology. Technology ought to adapt to people, not people to technology. For example, it is not without reason that we appreciate user-friendly tools. That is the way all tools should function. When that is the case those who operate them will also have greater pleasure in doing so.

In honoring the norm of *justice* we oppose any and all injustice that the development of technology may bring about. Engineers, instructors, and employees must all ask themselves whether their contribution to technology does justice to the plant and animal kingdoms, to our sources for raw materials, to consumers, to society, to culture, to third world countries, and the like. This norm of justice is an intrinsic part of technology. When it is disregarded, the government must take specific measures to restore justice. It is worth emphasizing that the positive influence of technological developments that obey this norm will also be felt in the many sectors of society in which technology is playing an increasingly important role. That is especially true, for example, in modern agricultural practices and in health care. The crises in these areas can take a turn for the better when we maintain this normative sense of justice consistently.

All of the norms indicated above are opened up and deepened when people hold themselves to the ethical norm of *care and love*. We are called upon to nurture a concern and compassion for everything that has to do with technology. This, of course, includes caring for and loving our neighbors, far off and close by, but also protecting the great diversity of God's other creatures. When love binds itself only to scientific-technological control, people become so obsessed with control that it is frightening. When this norm is not honored on all sides, people become more and more alienated from their work, for example, the farmer from the land, nature, and his animals.

The last norm to which scientific-technological development is subject is the *pistical norm*, that is, the norm of faith. The norm for faith, in a restrictive sense, is trust. Those who use the tools of technology have to be able to trust that these tools work and are safe. When this norm is met, they will be. Or more broadly, when people accept and operate according to the framework of norms sketched above, technology will be safe. That does not necessarily mean that all of technology's problems and hazards will be completely abated, but that these problems will be kept within bounds. We have to remember that with all cultural work there is a price to pay. No cultural work is a "path of roses." "Thorns and thistles" will continue to accompany the works of our hands. Within the integral framework of norms that we have sketched, however, cultural problems will not become intolerable or insurmountable. That is due to the transcendental direction of the norm of faith, namely, the acknowledgement that nothing in the world, as creature, stands on its own, but is dependent upon the Creator and ought to be directed to Him. This direction is the meaning and fulfillment of everything in the Kingdom of God. When people who trust in technicism and believe in the scientific-technological mastery of all our problems, self-confidently think to put themselves in the driver's seat, the ultimate effect will be the very opposite of what they expect. The problems will expand into dangerous, intolerable tensions, some of which could easily erupt into catastrophes. This faithful allegiance to technology is the background for the problems that technologizing calls forth (see §22).

Before concluding this paragraph I must mention the context in which technological science and modern technology is developing. The normative framework sketched above holds as well for a "free market economy," assuming one takes free in the sense of freedom within responsibility. In reality, however, economic forces often bolster technicism and the process of technologizing. That is why many turn to politics. But a materialistic political program will advance the same process. And yet, because the workings of government affect everyone and legislation can place restrictions on business and industry, the political arena is the place to address and deal with developments that have gone awry. This political activity must be attuned to the normative framework in general and, given the nature of the political realm, focus particularly on questions of fairness and public justice. Presently pressing problems will require that we take steps to contain the process of technologizing. It is possible to make a political choice for a normed technology that has a more sensitive attitude toward nature, animals, and the environment and still does not decrease employment. We have to realize that this kind of national political program can be effective only if this sense of equity and public justice also finds political support at the international level.

25. Technological Science Serving the Meaning of Technology

Technicism is one expression of the professed autonomy of humankind. Supplemented by modern science, it indisputably permeates our "technological culture" and, with that, so many different sectors of society. Industry, agriculture, and health care, to name just a few, all experience the pressure to technologize their ways. It is here that the problems and perils of modern culture are concentrated.

Technicism and technologizing, along with the reductionism that accompanies it, can be overcome through a renewed orientation toward the authentic dynamic of culture and, within that context, by explicitly posing the question as to the meaning of things. Technological science ought not to dominate modern technology. Technological science ought to be a productive detour that helps to nourish the wisdom of the engineer. With wisdom's discernment and circumspective insight, the engineer will be able to work cautiously and creatively, both as a scientist and as a technician, to advance the cause of a responsible technological science and technology.

Technological science and the scientific design method ought to satisfy the standards set by the normative framework sketched above.

Following this ethically acceptable path will contribute to a meaningful perspective for the development of technological science. Technological science will then serve the meaning of technology.

The meaning of technology in the perspective we are sketching is exceptionally rich and can hardly be articulated comprehensively. But it does include at least the following. Technology can alleviate in part the bind in which humankind naturally finds itself. Technology can increase life's possibilities, decrease physical burdens and difficulties at work, and free people from routine activities while opening the door to all kinds of mental and creative labor. Natural disasters can be averted, illnesses overcome, and, in a certain sense, with the aid of electronics and microprocessors, the deaf can hear again, the blind see, and the lame can walk again. Technological development can provide houses and food, supply a degree of social security, and increase available information so as to extend and deepen communication. Greater harmony between technology and nature is possible. Through all of this the responsibility of humankind grows as well. Material prosperity will not have a stranglehold or gain the upper hand, if it keeps in step with mental and spiritual well-being. The many gifts and diverse qualities of individuals and peoples will have a chance within technology and by its means. When it is situated within the perspective of an integral framework of norms that holds for all cultural activity and its hazards are kept within bounds, technology will make room for re-creative activities and a rich cultural involvement that are in balance with a conscientious stewardship of nature.

In this meaning-filled perspective technological science, the scientific design method, and technology will be able to contribute meaningfully to other cultural sectors, such as economics, agriculture, and health care.

Chapter Four

CURRENT ISSUES

26. Information Technology

George Orwell, writing in 1948, shared his glimpse of the future in his fascinating novel *1984*. His projections uncannily resemble our information age. "Big brother is watching you!" could easily become a reality in an age defined by the permeating presence of communication and information technologies. So many things, like production processes, social services, and even political activities, have changed drastically through the use of the computer. Will an "electronic shadow" soon accompany us from cradle to grave? By connecting new and existing information systems, we can monitor, control, and possibly even manipulate everyone registered in the system.

Dissatisfaction with this development is growing. Computer prowess has caused a sense of impotence among human beings. Genetic engineering, as well, opens a vista of imponderables. Technology's attempt to streamline society perfectly is accompanied by human alienation and societal disintegration and aimlessness. This chapter will focus on these current issues.

a. Optimism. Curiously, whenever computer experts discuss the meaning of computer technology for society, they present an optimistic and generally simplified view of the future. The standard version has it that we are entering an era in which all human activities will be controlled by electronic information systems. This will result in the following changes: two-way television, electronic banking systems, computerized education, electronic news media, electronic mail service, shopping by computer, and the like. The industrial society, in which people depend on the production of materials, will change to an information society that will satisfy the economic and social needs of all. While current society may be characterized by the division between rich and poor, between the privileged and underprivileged, some claim that the new information society will obliterate all these differences. Everyone will have instant access to enormous quantities of valuable information. Whereas other areas of life are plagued by limitations to growth, electronic technology and telecommunications have no such restrictions. Access to these services is becoming increasingly cheaper and more manageable. There is virtually no limit to our consuming and producing information. As people are liberated from the need to work, some claim that alienation will vanish and be replaced by a more genuinely human environment that promises cultural growth. For the first time in history, concentrations of power and social hierarchies will disappear. A resurrection of democratic ideals, better techniques for controlling the future, computerized education, and a global boost in human creativity will fulfill the lofty promises of a global village.

b. The Technology of Hope. Why this optimism? After all, the problems and faults of industrial technology have actually caused widespread pessimism. Nevertheless, belief in the idea of progress through modern technology is still pronounced. Whenever new developments are introduced, such as DNA technology, laser technology, or information technology, people's hopes for technological redemption are rekindled. The experts always promote their new technologies as solutions for age-old problems and as well as for those problems that resulted from the now largely obsolete industrial technology.

Obviously, this worship of technology is really nothing but a repristination of the old idea of progress. New inventions often have this effect on many people, so much so in fact that the new information technology is, with no proof, widely believed to be capable of producing a better world.

Minski, a prominent computer expert, when discussing fifth generation supercomputers, remarked that the new freedom will enable everyone to organize his life exactly as he wishes. Apparently, self-realization is now within our reach. A religious renaissance is upon us that features, instead of faith in a supernatural God, faith in collective human wisdom and the collective human spirit. At long last the earth will become a place of peace and rest, if only we buttress our attempts to develop information technology and make it available to all. The hope that the newest technology will usher in a glorious future, together with the continued blind belief in the idea of progress via technology, would like to keep us from looking critically at this technology. The old lesson that unnormed developments are nearly always accompanied by unfavorable side-effects and wholly unexpected problems is completely ignored. The experts are by and large blind to the real dangers associated with information technology. As a result, they excuse themselves from seeking responsible solutions to the most pressing problems.

c. Computer democracy. The computer expert Licklider supposes that by using information and communication technology can rekindle democracy. Such was not possible in so-called mass cultures. But today, says Licklider, by applying the newest techniques, millions of people will be able to discuss issues with each other and then decide democratically by simply pushing a button. The dreams of many social reformers will finally be realized. Since the available information will be universally accessible, everyone will have equal political influence and power. In other words, true democracy will be fulfilled in a computer democracy.

The first flaw in Licklider's view is that he sees democracy as first of all a matter of distributing information. In fact, the information will always be interpreted and selected on criteria or principles not necessarily shared by all. Because programmers use different criteria than the users, certain information may be withheld. Moreover, unlimited information may not raise political awareness; the "technologizing of information" can have an alienating effect. An overabundance of information can result in incoherence and disorientation. Unable to see the woods because of the trees, many people will be tempted to give up in apathy.

And what about the plans that are presented to the voters in the first place? They originate with a select group of researchers and planners of the future who determine the format and content of these plans and thus retain a good deal of control. Thus the people's freedom of choice will be restricted to the preconceived options of the experts. This is all the more true when communication media present a particular proposal as self-evident.

The delusion of computer democracy becomes even more transparent when experts tell us that decisions made through such a direct system of democracy can be predicted by data analysis and can thus be incorporated into the new plans for the future being proposed. Computer simulation will predict psychological and social reactions to proposed social changes. With this information in hand, computer technology will supposedly be all the more capable of designing and administering a better future. Current technocratic tendencies are thus reinforced by this "computerocracy."

d. A Computerized Future. Computer literature commonly suggests the notion that computer technology can better engineer and control the future than can human beings. They are considered incapable of absorbing all the information necessary to control the future. However, with a computer, different kinds of models of the future can be simulated and predicted precisely, allowing one to choose from among the various models thus constructed. The working out of such a model is then left to the devices of the expert. The basic idea is that we can shape a future that is in line with our wishes, solve all the eventual problems, and avoid all future threats.

However, attempts to create a global model of the future that integrates every possible factor have not proven to be very valuable. The models designed by the Club of Rome two decades ago have proven to be obsolete. The model called "Global 2000," which appeared in 1980, now seems to be largely irrelevant and needs to be supplanted by another model, "Global 2000 Revised."

World history cannot be programmed into a computer model. We cannot presume to have a complete overview of the future. Insofar as we do have some insight into the factors that influence the future, we still do not understand the scope and degree of the interaction among the various factors. The future is really inscrutable and full of surprises and upheavals. But if we continue to pretend that information technology can conquer the problems of the future, we will only find those problems reappearing in a more intense form.

Some experts will seek to carry the continuity of the past through the present into the future. But this tactic will only reinforce the problems of the past, especially humankind's loss of freedom. In reaction to this loss of freedom, other people will actively seek to oppose computer technology. They will seek to liberate their peers from the computer through revolution. As a result, computer technology will aggravate the old tension between power and freedom. An information society without tensions will not happen.

To avoid misunderstanding, I should point out that computer simulations and models of the future can help map out courses of development. But such a mapping-out has principal limitations. No one should assume that these projections can serve as blueprints for the future. Such maps are for reconnaissance purposes only, not for plotting a course for humankind. There must be room for human freedom and human responsibility. Computer simulations should serve, not curb, such human responsibility. Information technology must assist people in their quest for responsible courses of action in the future.

e. Computerized Education. Publications about computers in education are voluminous. Experts cherish high hopes for computer aided education. Predictably, commercial enterprises overestimate the computer's value in education. That education will improve through using information systems is not a foregone conclusion. Such supposedly self-evident conclusions are dangerous. Whatever comes to us via information systems results from scientific analysis and the scientific-technological control of information. But science works through abstractions. The methodology of science, as we saw, reduces reality. This means that the scientific information programmed in computer systems is always incomplete. Moreover, since scientific knowledge is of a general, universal nature, the information is also universal. In a very real sense, information technology presents a reduced, generalized image of the world.

When the computer dominates education, student users may adapt themselves uncritically to the computer's reduced worldview. Whatever does not make it to the computer screen is then simply not considered by the student. Computer programs contain only information that has to do with measurements, sizes, and statistics. Whatever falls beyond those measurable quantities—for example, faith, hope, and love—are considered unreal because they do not occur in the computer programming. Thus Minski errs tremendously when he equates computer information with wisdom. Computer information represents a specific form of human knowledge: it is scientific knowledge programmable in mathematical-logical formulas. It can never contain all forms of human knowledge, such as a knowledge of love, a knowledge of practical reality, or a knowledge of trust. True wisdom, which begins with the fear of the Lord, cannot be discovered by a computer program.

The type of education that seeks to transmit culture and to prepare young people to accept their responsibility for culture can certainly use the computer as a teaching aid. As a teaching aid, the computer can save an enormous amount of time. But the free time that results must not be filled up with other computerized activities. This would set students on a course of narrow specialization. The extra time should be filled by forms of education that cannot be accommodated by the computer. Thus, in addition to the growing specialization promoted by the computer, students must reflect more generally on reality. By pursuing both avenues, students may use the computer responsibly and wisely.

f. Computerized Labor. Many utopias sketch a future in which people no longer have to work. The world of our dreams seems to be a land in which we have to do nothing. And yet, now that we have liberated many people from their labor through automation and robots, we are beginning to realize how large a problem unemployment is. Recent economic recessions may be in part to blame, but that does not take away from the fact that when industry turns to contemplate new investments, it usually thinks in terms of automating production. This means that even if the economy picks up, technological developments and their new automated and robotic potentials will not necessarily create more jobs. The "utopia" of not working will follow.

Having nothing to do will not bring happiness because of the increased social problems and unemployment costs resulting from continued automation. Even if those problems could be solved, the general emptiness, boredom, and loss of human creativity and applied skills would continue to undermine utopian ideals because of the reduced worldview of scientific technology. If we see everything only in terms of science and technology, in the hope that it will somehow improve our lives, then the solutions can only be of a scientific-technological nature.

That in itself is an enormous reduction of what constitutes human happiness. People may often seem to be content to live by bread alone, but from a spiritual point of view, they will be impoverished. In the end, their material security will also be affected by science's reductions.

In short, the reduced worldview of scientific-technological control common to the information society confronts us with a caricature of the most beautiful promises ever uttered. The origin of that caricature is not to be found in information technology as such, but rather in humankind's materialistic inclinations, which have already betrayed their emptiness and hopelessness.

g. Demythologizing the Information Technology. As with many revolutions, it can also be said of the computer revolution that it devours its own young. If we expect everything from information technology, we will be confronted with serious tensions and an appalling void. It would be easy, but equally dangerous, to take a reactionary stance. What we have to do is identify the proper place of computer technology. Instead of allowing it to render our responsibility superfluous, we have to realize that it increases our responsibility. Without human responsibility, an information society will develop almost automatically. Computer systems will evolve into more complex systems which reduce the user to a slave of the system.

To maintain the *service function* of the computer and to exercise *effective control* over it, we need to analyze the structure of information technology. For the moment, however, I will underscore the limitations of information technology and define responsible potentials within those limitations. This task could be described as demythologizing the myth of information technology.

This demythologizing begins by asking what is meant by information. Information means *news* within a specific situation, something that was previously unknown. This new information is something people will have to consider in their own situation. Every bit of information is related to their responsibility.

But the word "information" in the phrase "information technology" is in a certain sense misleading. The "information" of computer technology amounts to a *reduction* of information, as pointed out earlier. Though we must conform to conventional terminology, we must also realize that there really is no such thing as information technology. For what happens in computer technology? By analyzing and abstracting, we formalize the human information process so that we can program. We program data in a computer according to a specific method, in a certain way. The procedure is referred to as an algorithm. The computer has a technically shaped physical structure that absorbs data previously processed by mathematical, logical analysis.

What happens in the computer may be called *signal absorption*. The computer absorbs technically rated signals that have come about through a mathematical, logical programming of human information processes. The signal absorption of the computer reduces reality enormously. The myth in current information technology lies in the fact that experts simply will not recognize this reduction. But if we remain aware of this reduction, we can see both the potentials and limitations of information technology. Its potentials include the speed with which a computer can absorb and reconfigure data, a very helpful feature. The limitations of information technology should discourage people from seeing the computer anthropomorphically, as a kind of mechanical human being.

Strictly speaking, computers cannot even perform simple mathematical functions, such as adding and subtracting. It is people who use the numbers. Fully human means of thought will always remain somewhat of a mystery to strictly scientific analysis. Admittedly, scientific analysis can convert human calculations to computer programs, but such programs are technically rated signals that only represent numbers.

Similar considerations also lead us to question the idea that computers think or possess some sort of artificial intelligence. In a technical sense, the process of signal absorption has lately been expanded and improved greatly. We should not take lightly the potential of what we call information technology. But computers will never replace human beings. They may be able to help us solve our problems better. When used responsibly, the computer can positively contribute to politics, education, and business. Within the context of human responsibility, which is ultimately a matter of living *coram Deo*, we should recognize the dangers of a narrow, scientific-technological image of the world. With a sense of responsibility we can use meaningfully the limited—and therefore potentially far-reaching—possibilities offered by computer technology.

h. Responsible Use of the Computer. Earlier I noted that the current trend in culture proceeds from systems analysis by way of systems designs to system control. In this way, the computer is incorporated whenever possible, and technocracy is perfected in computerocracy. Modern culture becomes a *Babel culture* (see Schuurman, 1987) with a *universal* language. It may appear that the problems of technocracy can be solved through systems philosophy and computer technology, but in fact they are only temporarily disguised and will shortly manifest themselves more intensely. This is true because as new avenues continue to be opened up to systems analysis and computer technology, these new avenues are also controlled by the idea that consistent application of these procedures is necessarily good. This process has contributed to increased alienation which in turn fosters such phenomena as unemployment, loneliness, isolation and standardization. For if computers are used wherever they can be used, many social contacts, human interrelationships, and opportunities for mutual sympathy and creativity are lost. If, in addition, we attempt to solve human and social problems by applying systems philosophy and computer technology, we will see those problems reduced to simple, technical definitions to be attacked with equally simple, technical solutions.

But it should be clear by now that the search for technological solutions to human and social problems will only intensify these problems. Dehumanization will result. In the meantime, computer technology appears to have become a nearly autonomous force. Enormous unmanageable stores of information are combined into a strictly centralized system. Information systems now operate autonomously selecting and interpreting whatever relevant information is necessary to make important decisions. One result of all of this is that people feel that their own responsibility is greatly diminished.

I have emphasized our need to relativize this systems thinking with respect to the original, primary world of experience and our need to emphasize human, creative responsibility in the relationship between abstract theory and full, concrete, practical reality. It is clear that computers should be used critically and cautiously. Wherever computers are used, it is essential to discover what kind of world-and-life view lies behind their use, which hidden philosophies are prevalent, and which viewpoints on science and technology are operative. The findings will determine whether we use or do not use computer technology in a given situation. We must also remember that constructive computer use can itself decline over time. For example, the collection of personal data for social care could in time begin to invite invasion of privacy.

Computers can be used responsibly in many ways. The Love Command requires that we explore computer possibilities for helping the handicapped. Furthermore, after admitting the computer's limitations, we should also use it in education. And we can use computers wherever the norm of efficiency is the highest consideration. It certainly has a place in space travel, but also in heavy, dirty, hazardous, and boring work. Computers can warn us about environmental pollution;

they can also help us use energy resources and raw materials more economically. In many current cases, however, it is not at all certain whether using a computer is indeed suitable. In such situations, we must consider the permanent responsibility of the person who introduces and uses the computer. In all cases, the user should be able to explain why a computer is necessary and for what purposes it is to be used. In a certain sense, this calls for a sense of social responsibility, the parameters of which should be defined by the government. Furthermore, incomprehensible computer programs should be made comprehensible to prevent users from transferring their responsibility to the computer. And just as a single user should not transfer responsibility to computers, the same holds for other individuals or groups of people. They, too, may not be robbed of their responsibility.

Only if people are prepared to live responsibly, remembering the purpose of computers and rejecting the temptation to exploit the computer in various kinds of technological power trips, will it be possible to expect a meaningful future with unprecedented and challenging possibilities. To recapitulate: systems thinking and computer technology do not result in their own unavoidable, autonomous development. Within normative boundaries, computer technology, together with an adequate sense of personal and social responsibility, can contribute much to a decentralized, pluriform, stable, healthy, and enriched culture. Certainly, much work still needs to be done. That work can be accomplished only if undergirded by a conversion of the scientific and technological models of life and culture. In those models, the real meaning of science and technology has been compromised. To see again the meaning of creation as the Kingdom of God is a challenging mandate. That Kingdom perspective does not exclude the computer, but instead makes it serve Kingdom activities.

27. What is Permissible? Thoughts on Genetic Engineering

A legitimate question in technology is whether it is permissible to do everything that is possible. This question is especially crucial in modern technology, based as it is in the natural sciences. This question was also asked when the first steam locomotive came upon the scene. Many people protested strongly, for they thought that the noise caused by the locomotive would prevent their chickens from laying eggs and their cows from producing milk.

This example has been used frequently to show that our ethical reflections on such matters can be flawed. The implied conclusion is that any new scientific knowledge permits its application. Yet the ethical problems can never be completely suppressed. One obvious example is the practical application of nuclear physics. Both the development of nuclear weapons and the technology of atoms for peace beg the question of our responsibility in managing the new scientific possibilities and technological achievements. Generally, however, ethical considerations are thought to be something like an afterbirth, and just as irrelevant. Space explorations continue unabated, and few people wonder whether the money spent on these developments is money spent responsibly, especially when one contemplates the enormous global problems of hunger and poverty.

Of late, ethical questions have once again moved to center stage, in the wake of the rapid advance of new technologies. Yet at the same time, the development continues unabated. The gynecologist Dr. Haspels recently observed that the time for ethical reflection on test-tube baby technology is long past. Since the development is now going full steam ahead, he concludes, it can no longer be stopped.

I wish to raise the matter of responsibility concerning genetic engineering. Of special importance is the recombination of genetic information—the so-called recombinant DNA—a matter related to test-tube babies and artificial insemination. The new discoveries of recombinant DNA technology increase the possibilities for artificial insemination.

a. The Experiment. At present, genetic engineering, the exchange of genetic information among various species, is still largely in the experimental stage. Many scientific researchers feel that their research has nothing to do with ethical problems. Ethical questions, they claim, would arise only when this new form of biotechnology is applied to industry, agriculture, animal husbandry, or medicine. Basically, they argue, scientific research aims at obtaining new knowledge; ethical issues belong only in applied science.

But the nature of the experiments in such pure scientific research shows that this approach is flawed. Besides, pure scientific research is the first step, whether large or small, to concrete application. Thus ethical questions must be raised at this point, if only to avoid the later excuse that ethical questions are no longer relevant since this research has already begun anyway.

First, we need to look at the nature of the experiment. Ever since Galileo, the experiment has served as a powerful tool for obtaining scientific knowledge. By experimenting, we find answers to scientific questions. The experiment actually simulates a small piece of reality. In doing the experiment, the researcher starts with his knowledge about given reality, but then proceeds to change it to the extent that the experiment's conclusions will provide him with uniform answers to his abstract, scientific hypotheses.

Most of us have had some experience with this form of experimentation in high school. There are few objections to this. The picture changes, however, when we begin to carry out our experiments on animals. Many people consider vivisection to be ethically irresponsible. In fact, legislation now restricts experimentation with animals to very specific conditions. Oddly, there is no similar legislation restricting experimentation with people, even though the introduction of such legislation is expected soon. The fact that such legislation is necessary indicates that experimenting with people is on the increase. In any case, it is evident that biology and medicine raise more ethical problems than physics does. Many people consider experimentation simply necessary to promote the high ideal of new scientific knowledge. It may, at first, be quite harmless. However, advanced scientific research requires more precise and more incisive experiments to address the increasingly more complex scientific questions. Not even physics escapes this dilemma. For example, when experiments are conducted with radioactive material, the ethical problems can no longer be ignored. But when biology begins to experiment on living beings, the ethical problems are immediate and should be addressed at once.

This can be illustrated from the field of molecular biology. To gain a greater understanding of how living beings function, scientists now conduct large-scale investigations into the structure of the DNA molecule, the basic building block of information in all living organisms. Recently, molecular biologists have developed a specific experimental method to find out more about the structure and function of the DNA molecule. Sections of a DNA molecule from a particular species are introduced into the DNA molecule of a bacteria. These experiments do not involve pathogenic bacteria, but rather bacteria that are harmless to humans, animals, and the environment. Bacteria are used because of their rapid growth: they are capable of almost instantly producing the desired changes.

As was noted previously (§19), American scientists in 1974 decided to halt such experiments because they could produce bacteria which, although not immediately pathogenic to the carrier, could readily become so. The freeze did not last long. Subsequent experiments were subjected to a number of security regulations, but since experience proved the immediate concern unfounded, the rules governing these experiments were gradually relaxed. However, caution is still in order because it is impossible to determine beforehand whether such experiments with bacteria are entirely harmless.

Needless to say, the new knowledge obtained from genetic research has been used. This knowledge enables scientists to rearrange the information in genes or to modify and manipulate it. Seldom do they see serious ethical problems with combining components of genetic information of a higher species with those of a lower species, such as bacteria or spores, provided the proper precautions are observed.

Combining the DNA of a certain plant species with the DNA of certain animal species offers us many opportunities, I believe, as long as the possible negative effects are considered right from the start. In fact, I think it would be irresponsible to disregard the potential benefits of this development. It could solve many of our present problems. To reject this technology could be more harmful than to use it judiciously. Outright rejection of this technology is often a reaction to claims that genetic engineering can enrich certain animals—or even make them almost human—and may well dehumanize people. A common argument against the recombination of genetic information from plants and animals is that this science threatens to erase the barriers between species. Aside from the question of whether this formulation is correct, I would maintain that recombinant DNA technology does not deal with the barriers between species. The introduction of a small segment of DNA does not create a new species. Admittedly, new characteristics are added to a given species. But this kind of "species modification" has, quite rightly, long been standard practice in agriculture and in animal husbandry.

b. Applications. Recombinant DNA technology has led to large-scale applications, many of which have been very constructive, the ethical questions not-withstanding. A recent committee report dealing specifically with this biotechnology shows how the recombination of DNA has led to new developments in pharmacology, agriculture, animal husbandry, medicine, and various industrial processes.

The insulin now used to treat diabetes is a breakthrough resulting from biotechnology. It is possible to produce various types of interferon which are claimed to have some effect on cancer. Similar combinations have led to the discovery of drugs to combat foot-and-mouth disease. It is no longer necessary to isolate these drugs from harmful viruses. They are produced by bacteria that have been infused with segments of alien DNA.

To a certain extent this new technology can remove the problems and dangers associated with producing synthetic chemicals. Recombinant DNA technology is also thought capable of producing bacteria that can absorb oil and degrade heavy metals. This would aid those whose responsibility it is to clean up the environment they have polluted. Science is currently working to "construct" plants that can absorb nitrogen directly from the air. This could make fertilizer more economical and more plentiful, thus helping to relieve hunger in the world.

Many of these proposals are still only daydreams. Much research remains to be done, especially on the potential negative effects of the industrial production of these bacteria. Another relevant ethical question is whether it is responsible to introduce newly created bacteria on a large scale into the environment. Will they not upset the delicate balance in nature and, if so, to what end? Some people consider this to be such a critical question that they strenuously resist the planned bioreactors and the general introduction of recombined bacteria into the environment, with the same energy that others protest nuclear reactors.

Indeed, we must be cautious. Current recombinant DNA research is justifiable only if it conforms to the safety regulations. Large-scale, indiscriminate application of these techniques will unquestionably have some harmful side effects. It may lead to the production of new biological weapons. But even with peaceful intent, we cannot ignore the fact that we have no experience at all with many of these plans. It is impossible to anticipate what will happen. Current industrial technology teaches that there will always be negative results that could not have been foreseen. We must heed the lesson. It is too optimistic to expect that any new technology will automatically cure the problems created by other technologies.

c. The Scientist's Responsibility. Because of the problems and dangers associated with experimentation and its industrial application, and also in view of the possible wrong use of what may otherwise be sound scientific knowledge, the responsibility of the scientist has once again become our question. Such questions have been asked before. For example, was the nuclear physicist Otto Hann responsible for Hiroshima? Are nuclear physicists responsible for the development of nuclear weapons systems which, because they are becoming less manageable from a technical point of view, could lead to demonic catastrophes? Obviously, such questions must be asked of scientific research leading to any new technological breakthroughs.

As pointed out earlier, recombining genetic material can have both positive and negative effects. Are researchers responsible for the effects? Some say they are responsible only for their own research. But how much greater would their responsibility be if the results of their research in DNA technology were to cause irreversible harm to humans, to animals, or to the environment!

Others rightly object to this kind of reasoning by noting that the harmful effects would be much greater if we halted all biotechnology altogether. World hunger, for example, would be much worse.

In view of this ambiguous situation, the question of the researcher's responsibility is not an easy one to answer. We are becoming aware that scientists must be very careful when introducing new technologies. Certainly, scientists and their support organizations must accept more responsibility. Yet the form of this responsibility, the way in which it can be discharged, is one of the most crucial problems facing the scientific community and, to a large extent, the political community as well.

28. The Genetic Engineering of Human Beings

We turn to ethical questions concerning the potential of genetic engineering as it applies to man. Either directly or indirectly, recombinant DNA technology also enables the scientist to engineer human genetic information. Are there ethical limits to this power?

Charles Weissmann, Swiss molecular biologist, once said that in the near future every person will be able to apprise himself of the nature, structure, and function of his own genes. A gene is the smallest unit of hereditary information. The recombinant DNA enables the scientist to map our hereditary information. That information consists of roughly one hundred thousand genes, ten thousand of which have thus far been researched. Before long, it is thought, scientists will completely understand our total genetic being. Aided by the computer, such "gene mapping" will become a manageable research procedure.

If such gene profiles are not protected, they can easily be abused, as, for example, through discrimination in the workplace and by prejudicial conditions made by insurance policies if the gene profile shows that the insured poses a high risk to the company. Naturally, such ethical questions become even more crucial when a reading of parents' gene profiles in combination with the prenatal investigation of the fetus' gene profile shows that the child has an "unfavorable" gene structure. Such research could easily lead to a higher rate of abortions. Going even further, Weissmann explains: "Deep down I know that at some point in the future—I speak of decades—it will be possible to bring about complex genetic changes in man. Then we will have to confront head-on the associated ethical problems."

We may not be in such a situation yet, but we are on the threshold of an age in which we will be able to manipulate procreation outside the mother's womb. Parents availing themselves of homologous insemination, can now be helped in the conception of their own child. Through heterologous insemination, it is possible to conceive a child whose father remains anonymous. The possibilities of the test-tube technique will increase to the point that fertilized ova of preselected parents can be implanted in other women. With the aid of test-tube technology, it will be possible not only to signal hereditary defects in advance, but also to "construct" a super-baby with the most favorable genetic structure. Thus the recombinant DNA technology will be applied, by way of the test-tube, to the human embryo. Clearly we are here presented with significant ethical questions. Some people rightly warn that these techniques have already claimed the lives of many fetuses that were not considered suitable for implantation or that were simply superfluous. Others maintain that the sole criterion for application of this technology should be the quality, not the viability, of individual life.

The available literature that advocates this so-called high quality of life also promotes the "creation" of large numbers of such high quality individuals. This is known as cloning. The technique of cloning was proven successful in mice and frogs, but has not yet been applied to human beings. But even if it were possible to apply it to human beings, would it be permissible?

I wish to discuss yet another experiment. In a certain area of research, scientists have tried to fertilize the ovum of a hamster with the sperm of a male human being. The result was a hybrid, a life-form somewhere between that of a human being and a hamster, that could develop to the stage of a few undifferentiated cells. Generally, experimentation with life forms between human beings and animals is not tolerated. Nonchristians appeal to the norm of human worth or humanity. Many nonchristians use this norm to argue for the prohibition of experiments that could lead to further experiments involving the different life forms. In borderline cases, however, objections are less strenuous. In November 1982, the British medical research council established a number of guidelines for such experiments. One such guideline would halt fertility tests that could lead to "the human-hamster hybrid system" before cell division started.

Whereas these tests already signal the beginning of hybridization, they, too, should be rejected. This is true especially since these tests could signal the beginning of genetic engineering of those varied cells which include the hereditary information of the human. By way of such tests, the technology of recombining genes could easily be deflected into new, and in my estimation, illegitimate avenues. In the meantime, scientists should look for more ethical types of experiments to test the fertility of the ovum.

I do not expect the resistance to such experiments to stiffen. Rather, because of the positive potential of test-tube techniques, the experiments are likely to increase. One reason for this laissez-faire attitude can be found in the widely accepted legalization of abortion. To a large degree, the legalization of abortion has promoted experimentation with the human genetic structure. The American ban on federal funding for the transplantation of tissue from aborted fetuses into humans has recently been lifted. Hard and fast rules governing these experiments can no longer be established. Similarly, nothing is done about experimenting with aborted fetuses kept alive for research purposes. Recent research journals claim that experiments with aborted fetuses are necessary to test new medicines that can be of benefit to humankind, for example, in the struggle against Parkinson's disease.

a. Objections. My objections against possible genetic engineering of fertilized ova in vitro and against manipulating forms of life between human beings and animals, including life-forms in their initial stages, are based on my rejecting the current view of humans that has long been subliminally accepted, but that has now found its way into the public domain through labs and hospitals.

Increasingly, people today are accepting and promoting a view of humanity that is thoroughly materialistic, a view inspired by the natural sciences and modern technology. It is said that human beings are machines, albeit very complicated ones, or a complex physical-chemical substance or, more recently still, an information programming system of complex DNA molecules. When one's view of human beings is tied to a specific dominant area of science, that particular view of human beings becomes the dominant one. In this way, specific, abstract views of human beings are absolutized. Their totality and unity becomes subservient to the theories of science and technology. The end result is a reductionist view of the human race. Having fallen this far, people can easily fall prey to the manipulation of other theories.

This reduced view of human beings was originally limited to scientific and philosophic circles. However, when the materialistic world-and-life view gained ground and the Christian world-and-life view went into retreat, this reductionistic view became commonplace. This secularizing of culture contributed to the fact that this thinking about human beings as purely material beings could be applied as theory more and more. Humans were dehumanized and compared to a thing. Fortunately, quite a few humanists also protested this dehumanization, since it challenges our humanity and human worth.

The materialistic view did not first become evident in genetic engineering. The modern way of using human resources in the labor process, as well as attempts to legalize abortion, have been clear evidence of this reduced view of human beings. Through the potentials promised by genetic engineering and the test-tube technique, the materialistic view has reached its zenith.

The Christian sees the individual, though corrupted and defaced, as the created image of God. God's love and care continually reach out to the individual. Therefore, we must love every one of our fellow human beings. Out of this love we can use scientific-technological possibilities to combat sickness and relieve suffering, as is done in medical science. But the Christian's view of human beings is based on the inviolability and integrity of every individual, as well as on the realization that we can only moderate human suffering and can never entirely repair the effects of his corruption. Viewed from the perspective of creation and fall, the origin of this corruption lies in human beings willfully breaking their communion with God. This religious failure cannot be remedied by science and technology.

As indicated, the materialistic view of human beings does not accept the inviolability of human life. Instead, it focuses on the *quality* of life. The purpose of technology is to obtain the highest possible quality. In this view, our broken life may in theory appear to have been repaired by science and technology. In a real sense, genetic engineering then also has to guarantee the quality of its product. Needless to say, a poor product is not acceptable and must therefore be destroyed.

b. The Test-Tube Baby. Rejection of the test-tube technology invariably brings the counterargument that the method is ideal for helping childless parents who have found the more natural avenues of conception closed to them. This is a noteworthy argument, provided the method of test-tube technology is not used unless absolutely necessary. Yet closer inspection and a thorough acquaintance with the literature available on the subject will show that more is involved here than may at first appear.

The many questions and problems of test tube technology can be avoided by helping prospective parents in other ways. I have in mind the Craft method which involves the removal of ova from blocked fallopian tubes and their implantation into the womb, together with male sperm, *before fertilization takes place*. This method can be applied on a large scale and is socially more acceptable because its costs are much smaller and thus its opportunities accrue not only to a small, privileged elite. From an ethical point of view, I believe this method to be acceptable, for the method does not allow for unjustified experiments and, secondly, the womb is where conception takes place and the embryo grows.

By contrast, if fertilization *takes place in the lab*, it is not possible to control the number and types of experiments conducted on ova. As well, it may mark the beginning of the recombination or manipulation of genes. As far as I know, no efforts have been made thus far to try to change human genetic information. However, the laboratory method of artificial insemination does require multiple ova, only one of which will ultimately be reimplanted. In the event of failure, any one of the remaining embryos will take the reject's place. Moreover, the criteria of artificial insemination demand that the ultimate product meets certain standards of excellence. Thus the test-tube technology itself must be able to guarantee the quality of life. This in turn means that selections must be made between products that are likely to be successful and those that are not, or perhaps less so. Is it up to us, however, to make such judgments and take decisive action? This can happen only if the individual is no longer seen as unique in the eyes of God and when human beings arbitrarily decide which human life meets current standards and which does not.

There is another equally important argument. From a number of publications, it appears that this technology is used to conduct uncontrolled scientific experiments on test-tube embryos. Through these experiments, researchers are encouraged to explore all kinds of new avenues. Any technology that offers such unlimited and uncontrolled opportunities to researchers must surely be rejected.

I previously mentioned that the test-tube technology has thus far not been used to bring about planned changes in human genetic information. However, this has been done with animals. Rat genes pertaining to the rat's growth hormones have been implanted in the embryos of mice. The mice so manipulated turned out to be super-mice, whose offspring showed evidence of the earlier gene transplant.

The successful experiments conducted on animals opens the way to similar experiments with the genetic information of human beings. The very possibility requires us to deal with it at this time. Repairing genetic defects, infusing into people specific characteristics, such as greater intelligence, no longer belongs to the area of science fiction. However, before this can be put into practice, many experiments will have to be conducted, and this necessary experimenting with human beings cannot be justified. In a recent article, Professor Leenen of the University of Amsterdam raised a number of questions to which there are no unambiguous answers: What possible dislocations can the change of genetic information have on an organism? What unknown characteristics will be brought on by the new combination of genes? What is the effect of reinforcing a particular characteristic in terms of the overall balance in an organism? What happens when a genetic experiment fails?

My conclusion is that test-tube technology will surrender the total human genetic structure to scientistic and technicistic models. For that reason, it is not at all certain whether the high expectations will materialize. Quite possibly, the opposite could be true.

As a matter of fact, the opposite is certain. Materialistic human beings now think they can offset the ill effects of what was a religious failure, the fall into sin, by means of science and technology. Since human beings, without God, are no longer capable of bearing or absorbing the effects of sin, such as suffering and evil, they seek to redeem themselves, but in the process invite more suffering and evil upon themselves.

Does this mean that all genetic engineering of human beings must be rejected? No, not if the supremacy of science and technology is brought low in order to be turned to service. Only then will there be an end to the scientistic and technicistic approach to humanity. Only then can we enlarge on the perspective of a normative development for science and technology. I can appreciate genetic engineering if it is applied to remedying certain hereditary deficiencies and diseases, but only if the approach zeroes in on limited genetic therapy on the *organic level* rather than on a general treatment of the *total genetic structure*. In any case, all the ethical rules of medical experiments must be strictly observed.

c. Conclusion. Objections have been raised against the use of the term test-tube baby, and rightly so, because the baby does not come from a test tube. But I hope that the term "test-tube" continues to serve as a reminder that conception takes place in a laboratory. Temporarily, the pregnancy is interrupted. That brief time is filled with human technologies, for which the recipients will want some form of guarantee ensuring the quality of the "product." Right from the beginning, the materialistic process has degraded human beings to things to be manipulated.

Even the intent of this technology—although there may be no plans to experiment or keep back some spare embryos—must be rejected because of its many unforeseeable consequences. This is all the more so since research into the Craft method has not come to a dead-end.

Craft's alternative method, however, does not get the attention it deserves, primarily because it precludes any tinkering with the embryo. Generally, support for experimentation comes from secularized religion, the atheistic religion of Western man, which was developed from the Renaissance, through Rationalism and the Enlightenment, to our increasingly scientistic and technicistic culture. The technology of the test tube baby and the many possible experiments associated with it show clearly the post-humanity of our technicistic age.

Someone might possibly object that, because ethics is always undergoing change, I should be more careful with my dismissal. That is often true, especially in a secularized culture where Christian norms and values, long cast in stone, have suddenly begun to shift like sand. However,

I would hope that our ethics will not change when it comes to conducting experiments on children or the mentally impaired. Why should our ethical position change with respect to embryos? My position makes clear where the boundaries are; and even then, much reflection is needed to prevent responsible technologies from degenerating into acts of unspeakable evil.

Without a doubt, the problems surrounding modern science and technology are becoming more menacing. The responsibility of scientists and researchers, together with all those people involved in various ways in technological development, is becoming steadily heavier and more difficult. But, as I pointed out earlier, that responsibility must also have a different orientation than the current one.

We cannot continue to dwell on symptoms, but must search out the causes of the various critical traits. We cannot limit ourselves to partial problems, but must attack the many challenges with an integrated strategy. The combination of science, technology, and secularization encourages solutions that may provide some short-term gains but will bring agony in the long run. If we elect to solve the problems caused by science and technology by applying more scientific and technological solutions, the process of escalating technicism will only be reinforced. If, furthermore, science and technology become poltical instruments, the result will be a technocracy in which we, though still making our contributions, will become the slave of anonymous, pseudo-independent forces.

I have pointed out in various ways the need to approach the question of responsibility differently from the way that is currently fashionable. In many respects, we will once again have to pose authentic questions, such as ones about the origin and meaning of things, about the coherence and diversity of everything, and especially questions about normativity for all of creation.

Christianity must accept this challenge. This means that the materialistic inclination, which in its diverse forms has reduced and sealed up a view of the world as a world without God, must make way for a profound, resolute, spiritual conviction. This conviction will attest to our dependence on Christ, our Redeemer from sin, guilt, and evil. It will also confess to our relationship to God, the Creator of the whole of reality. This renewal will offer a new direction to scientific and technological opportunities as it rejects autonomous, atheistic, and materialistic world-and-life views. In the latter, man remains the measure of all things, while in the Christian view, the law of God, as manifestation of God's will and summed up in the two love commandments, continues to provide direction to all things. Through it we can be liberated from the menacing superforce of science and technology.

Chapter Five

TECHNOLOGICAL CULTURE BETWEEN THE TIMES

—A Christian Philosophic Assessment of Contemporary Society—

29. Religious Groundmotives

When evaluating the development of technology in human history, Christians must look carefully at its spiritual-historical background. A number of Christians working in the tradition of Reformational Philosophy have focused on this religious-historical dimension. Herman Dooyeweerd, in particular, with his philosophy of the cosmonomic idea and his view of religious driving forces, or *groundmotives*, provides a good vantage point for assessing what is happening in our culture. This theory of religious motives needs to be elaborated in a more comprehensive reading of our culture.

Dooyeweerd originally introduced the notion of religious groundmotive in 1941 to expose and clarify the relation between religion and theoretical thought. In doing so, he provided valuable insight into the depth, diversity, and turmoil within Western philosophic thought. From the start he realized, but never really expanded on the fact, that every groundmotive, given its religious character, is related to culture. Religion, under the sway of one or another religious groundmotive, conditions not only philosophic thinking and science, it also influences culture in general and our scientific-technological culture in particular.

Technology is today the mass, the motor, and the mark of nearly every cultural activity or field. By focussing on how religion relates to culture, we can broaden and deepen our insight into what is going on in our culture, and into what we must do about it. Such a liberating perspective will help us to understand better the current crisis of our culture, also with respect to postmodernism.

Dooyeweerd identified three nonchristian groundmotives as moving three basic periods of history, namely: the pagan period; the period of synthesis between pagan thought patterns and biblical themes; and the humanistic period. The humanistic groundmotive, according to Dooyeweerd, originated in the pretension of human autonomy and is formulated as a dialectical relation between freedom and nature, whereby *freedom* is an ideal of the human personality and *nature* is the nonhuman world, scientifically conceived. Dooyeweerd originally spoke of science as a *religion*; later on he speaks only of the science *ideal* as an expression of human freedom. The dialectic lies in that freedom when absolutized creates an absolutized science that threatens the freedom that gave it credibility. He claims that this humanistic religious groundmotive has influenced the entire development of postmedieval Western culture.

Dooyeweerd understands religious groundmotives to be the central driving force in developing culture (1979), but surprisingly restricted his discussion of that cultural influence to the field of philosophic and scientific thought. Although these motives are radical and integral in character, he seldom examines what they imply for culture in general, perhaps because he deals with religious groundmotives, especially in his later work (e.g., 1953), in the context of his transcendental critique of theoretical thought. In that critique he argues that every philosophic position is grounded in a religious motive. This motive is a driving force arising from the heart as the religious center of the human being. But he says little, for example, about the structural consequences of the humanistic groundmotive in the development of culture. He does stress losing or reducing freedom in the dialectical process between an absolutized human personality and an absolutized science, but he ignored other cultural tendencies and problems that we are well acquainted with in our time. And so it is for us to answer the important question: How does the humanistic groundmotive influence the health and wholeness of our culture, a culture that philosophy and science influence greatly through technology?

Analyzing our technological culture in the light of the humanistic groundmotive will aid Christians and others who are seeking to find their bearings in modern culture. In this context we must also ask whether we should revise Dooyeweerd's formulation or description of that motive in light of the typically technological culture that has now evolved.

30. A Groundmotive Revised

Dooyeweerd spoke as early as 1939 about the groundmotive of "the control of nature" (see also 1960). "Control" here is not so much scientific as technological. There are advantages to speaking about a cultural qualification, but Dooyeweerd seems to apply this notion only to the field of philosophic and scientific thought. He wrote a great deal about the relation between Christian faith and science that is of interest to university scholars, but he did not broaden and deepen his analysis to include the influence of science on culture outside the university. For example, he says little about the influence of science on technology, about the influence of technology on science, or about cultural fields in which technology is a decisive factor, such as industry, economics, agriculture, health care, education, and politics. Certainly a groundmotive of technological control would leave its mark on developing science as well as, as is so often said, the other way around?

Historically, Dooyeweerd is right: the rise of modern science preceded the surprising rise of modern technology. But is the religious spirit of technological control not active earlier in both history and science? Dooyeweerd (1966) himself implicitly alludes to that spirit when he describes the religious groundmotive as a creation power attributed to theoretic thought which, after it has methodically broken down the God-given creation order, reconstructs an order according to the ideas of human autonomy. Although Dooyeweerd states that the humanistic motive of creative freedom is concentrated in theoretical thought, the substance of what he holds can also be interpreted to mean that the motive of creative freedom is concentrated in scientific-technological control.

Other examples of the same shift in accent can be mentioned: In the *New Critique* (1953) Dooyeweerd refers more than once to the process of rationalization in modern society. Yet when he speaks about the relation between science and technological progress, he clearly holds that modern technology is a consequence of science or scientific rationality. He does not seem to perceive that under the influence of the humanistic groundmotive, the relation as such between science and technology is distorted to the extent that the reverse is more plausible. Dooyeweerd's later formulations in his *Roots of Western Culture* (1979) indicate that he indeed was aware of that effect. He speaks there about the ideal of science as an ideal of control, as a technological ideal: "Proudly conscious of his autonomy and freedom, modern man saw 'nature' as an expansive arena for the explorations of his free personality, as a field of infinite possibilities in which the sovereignty of human personality must be revealed by a complete mastery of the phenomena of nature" (150). What Dooyeweerd is describing here is technology, or better: a technicistic spirit. In this light it seems clear that Dooyeweerd's original science-ideal can better be understood as the ideal of scientific-technological control. Such an ideal actualizes itself first in science and modern technology and subsequently in many fields of culture.

The content of the humanistic groundmotive can therefore better be formulated as the motive of the creative freedom of human beings—the ideal of personality!—concentrated in scientific-technological control or power.

If this is accurate, the consequences for understanding our culture are far-reaching. Given the guidance of this re-formulated humanistic groundmotive it would appear, for instance, that science in our culture is more often developed as applied technology than is technology as applied science. As early as 1936 Dooyeweerd in a certain sense already recognized that science is usually used as an instrument. Reality is brought under control with the help of scientific thinking. Instrumentalism here means subjecting creationally irreducible activities to absolutized technological control.

Henk Van Riessen expresses a similar view when he says in his *Society of the Future* (1957) that the crisis in our culture is caused by the spirit of absolutized technological power.

31. Faith, Science, and Technological Culture

In light of the above it is clear that our understanding of the relation between faith—as an expression of religion—and science deepens and broadens when extended to the relation between faith and technological culture—a process that may exceed the university context. In any case, we should always connect thinking about how faith relates to science with how faith relates to culture. Because we have not always done so in the past, Reformational Philosophy has contributed less than it might have to developing a normative perspective for modern culture. Now that our culture is in a profound crisis there is an opportunity to speak about this relation more pertinently than ever before.

Nowadays it is undeniable that the ideal of control has manifested itself in the history of culture as a disruptive force. The final result is, according to Dooyeweerd (1979), "God's judgment in world history." Today we see this historical judgment wherever people dehumanize fellow humans, destroy nature, pollute the environment, endanger lives with atomic energy, threaten nuclear war, and destabilize our highly developed technological culture.

The instrumental use of science leads to a reality shaped according to science, including its functionalism and universalism. Large-scale and unrelenting instrumental use of science's abstractions, with their inherent reductions, warps and will eventually destroy reality and its meaning. This loss of meaning is already evident in the fragmentation of nature and society. In bioindustries, for example, reproductive and production technologies often reduce the integrity or essence of animals to a mere economic utility. Furthermore, as a consequence of the fragmentation of global society a sense of harmony and social justice between the over- and underdeveloped countries is absent. And these are just some of the problems (see §7).

Efforts are made to solve problems by uncritically introducing new forms of science and high-technology, such as the systems approach, information technology (Laszlo, 1974), biotechnology, and even genetic manipulation. Under the influence of the humanistic groundmotive, every area of our culture has been technologized (Schuurman, 1987). It is seldom asked—and this is the critical question—whether the technology is suitable to all our problems, and especially to the problems technology itself has created, such as polluting the environment, deficiencies in agriculture, and so on.

32. The Dialectic of Culture and Nature

Reformulating one of the poles of the humanistic groundmotive—that of scientific-technological control—sheds light on the actual structural coherence and unity of many of our cultural problems. Insight into these problems gives us reason to reformulate the other pole of the humanistic groundmotive as well. Today the dialectic is oriented around (technological) culture and nature. *Nature* has come to mean *reality as organism*, and that is naturalism. As a result a dialectic rages between the anthropocentrism of the technological culture and the ecocentrism of a *counter-culture* committed to alternative technologies, alternative agriculture, and alternative medical care.

Thus the real content of the humanistic groundmotive of our time is that of *culture* and *nature*. Absolutized *technological control* enjoys primacy over absolutized *nature as organism* because no way can be found from the existing technological culture to a future ecological culture. Culture depends upon this control.

This dialectic sheds a light on our time that reveals it to be postmodern and at the same time neo-pagan. The orientation of many people to the pole of *nature* demonstrates the influence of neo-paganism in our secularized culture. This essentially pre-christian motive, which is connected with the religions of culture and nature associated with the Greek groundmotive of form and matter, has tended in the neo-pagan period of our time to deify either scientific-technological control (and often of the material welfare associated with it, as we shall see) or nature, "mother earth." The religion of nature, which is represented in several streams of the New Age movement, where the earth is adored as goddess Gaia, stands opposed to the religion of control, of technology. This dialectic characterizes the development of our culture. Such is the mind of our times (see §36).

Perhaps it is unnecessary to note that philosophers who are influenced by this groundmotive and who orient themselves to one of its poles dialectically manifest in their thinking the other pole as well. For instance, consistent environmentalists often speak of the "spaceship Earth." Some philosophers endeavor to achieve an impossible synthesis between the two poles (Bookchin, 1980 and 1982).

33. Technicism

So far I have tried to make clear that the crisis of our culture brought about by the humanistic religious groundmotive is not related first of all to science but rather to technology. Dooyeweerd concluded that scientism or rationalism is the dominant characteristic of our culture. Other representatives of Reformational Philosophy stress that economism is the main characteristic of the crisis of our culture. Bob Goudzwaard, for instance, in his book *Capitalism and Progress* (1980) locates the main characteristic of our culture in the *economism* of capitalism. Such an analysis is very fruitful. Ironically, however, in elucidating the reduction in modern economics, Goudzwaard more than once speaks about capitalism in categories of technology. Thus we have economic *mechanisms*, the *tunnel*-economy, the *spaceship*-economy, and so on. It appears that economism, too, is reductive and in a certain sense insufficiently broad and deep to make perfectly clear what is going on in our culture, and perhaps especially in our economy.

The Belgian philosopher Vermeersch speaks of the complex of Science-Technology-Capitalism as he seeks to understand our culture, but focuses his critique on science and especially on capitalism. Such a critique of culture is, of course, generally well accepted, at least among philosophers. Yet I want to stress that neither scientism nor economism but technicism (see §21) is the primary point of contact between the religious humanistic groundmotive and human activities. This is true because technology is historically, in the sense of classical technology, and ontologically prior to science (Ihde, 1985). It is technicism, or one can even speak of the implicit or hidden ideology of technology, that influences science and economy, because there seems to be no room for critical distance in their relation to technology (Pippin, 1994). *Technological push* has priority over *economical pull*. Science and the economy as such are usually interpreted technicistically, whereupon via positive feedback, they reinforce technicism. Then taken together they feed a greedy society.

There are other arguments too for giving priority to technicism. For example, technicism is the best formulation for the secularized motive of creation and redemption. The utopias of Robert Bacon in the thirteenth century and of Francis Bacon in the seventeenth century are already splendidly technological. Francis Bacon in his *Nova Atlantis* holds that we can overcome the consequences of the Fall by technology. Besides, history shows that economism, as materialistic economy, is not always all-determining, for instance during wars. Military technology or defense technology may require great financial sacrifices that have no positive effect on economic welfare. That happened, for instance, in the former Soviet Union. The technology of space travel cost a great deal to develop and reflected the conflict between the two superpowers. The competition between them was especially technologically qualified.

That technicism is much more important than scientism or economism becomes even more clear when we consider the matter of worldview. To speak of a "technological world picture or image" is more satisfying than to speak of a "scientific or economical world picture." Since the appearance of the theory of relativity and quantum physics, a scientific world picture offers no certainty and is fraught with many questions. The technological world picture seems to be stronger, broader, and reach deeper than an approach from scientism or economism (Dijksterhuis, 1986).

In short, the implicit ideology of technology, technicism, is the dominant expression of the humanistic groundmotive. Technicism entails the pretension of human autonomy to control the whole of reality: man as master seeks victory over the future; he is to have everything his way; he is to solve all the problems, including new problems caused by technicism; and to guarantee, as possible consequence, material progress. Technicism also always implies an obstacle or enemy: it may be God, nature, another country or state, or competitor. Isn't it amazing that technological development was the strongest during the Cold War?

Technicism reduces science to its instrumental use. The economy, as is obvious in Western culture today is also interpreted technicistically, with utilitarian economics as a complement. Moreover, the influence of technicism on technology itself is also negative. Technological development becomes a destructive cultural power.

What I have said thus far entails many cultural consequences. Technicism has left its mark on many cultural activities, namely, reducing, disturbing, and fragmenting such activities. The symptoms can be found not only in science and the economy but also in agriculture, in health care, in the form of instrumental justice, and even in ethics, where people today are talking about *ethical engineering* (Caplan, 1980). Even the Christian religion of many Americans has, according to Wuthnow (1988), been increasingly influenced by technicism. It is the spiritual background not only of large cultural problems but also of micro-ethical problems, including abortion, euthanasia, and procreation technology.

To make the picture more complete, I want to add that a good deal of theology, although often unknown to the theologians themselves, is also influenced by technicism, when, for instance, one speaks about God as a human fabrication or construct. Some philosophers as well, such as Marvin Minsky (1986) with his ideas about artificial intelligence, express technicism very clearly when they conceive of both society and human beings as expressions of a very complicated information machine or system.

Generally speaking, one can argue convincingly from a technicistic standpoint, that the main trend of Western philosophic thought is best characterized as *thinking through technology.* This means that science and rationality in general are distorted because they have been used as technical instruments in the service of technological power.

I believe that a great variety of cultural problems can be understood more satisfactorily from the standpoint of a critique of technicism than through other more common approaches and that we also get a better grasp of several irrationalistic streams by considering them as reactions to technicism rather than rationalism. Think, for instance, of existentialism, neo-marxism, counterculture philosophy, New Age thinking, and postmodernism. They all express the increasingly shared sense that we live in a ruined world of our own making.

34. The (Hidden) Ideology of Technology and Postmodernism

Postmodernism is a form of technological pessimism. That helps explain the controversy between postmodernism and the mainstream of the Enlightenment and its idea of progress. Postmodernism demonstrates the failure of Western technicistic culture and philosophy.

Although postmodernism proclaims the end of ideology, the ideology of technology is still at work in it implicitly. Therein lies the continuity of postmodernism with modernism. Leo Marx (1994) says of postmodernism:

> A common feature...of the umbrella concept of postmodernism, is the decisive role accorded to the new electronic communications technologies. The information or knowledge these technologies are able to generate and to disseminate is said to constitute a distinctively postmodern, increasingly dominant, form of capital, a "force of production," and in effect a new, dematerialized kind of power. This allegedly is the age of knowledge-based economies. (24)

Postmodernism is, in a way, the spirit and philosophy of the post-industrial society. Traditionally, power was thought of as entrenched. It could be attacked, removed, or replaced. But postmodernists like Jean-Francois Lyotard (1988) and Michel Foucault (Marx, 1994) envisage forms of power that have no central, single, fixed, distinctive, controllable locus. For the first time in history, concentrations of power and social hierarchies will disappear. An overabundance of information can result in incoherence, fragmentation, and disorientation. Thus it seems as if technicism is evolving from a central technocracy into an anarchic technocracy. Technological power is present everywhere but concentrated nowhere. Hence postmodernism acknowledges no normative direction for technology. It is *comfortable with change*. Micro-electronics, information technology, telecommunications, and systems technology seem to hold sway over history without a controlling subject and to alter the meaning of time and place. Everything is technologically possible and everything is technologically allowed.

This postmodernist outlook, when combined with the operation of multimedia, tends to validate the idea that life is dominated by large technological systems. Enormous, unmanageable stores of information appear to function autonomously in information systems. Computer programs become incomprehensible. As a result, the postmodernist attitude towards technology is one mainly of melancholy resignation or fatalism. Fatalistic pessimism is an ambivalent tribute, however, to the decisive, all-determining power of technology.

Even so, when postmodernists become active as technological anarchists, they manifest a senseless optimism about modern technology. The (hidden) ideology of technology in the postmodern era gives priority to the individualistic instead of to the collectivistic version of it. As such it expresses a new, postmodern form of dialectics in relation to technology.

35. Philosophers Today

The analysis of the absolutized culturally formative power of technology presented in this chapter is confirmed by several present-day philosophers from different backgrounds, each in his own way. Some promote the current development positively or optimistically while others are negative or pessimistic. One sees no way out, another tries to find a liberating or saving perspective. I have learned a good deal from several of them. While remaining faithful to my biblical, reformational basis and perspective, I have welcomed their insights with appreciation.

Consider briefly Heidegger's exposition. He holds that Western philosophy is already technicistic at its core as a result of its Greek origin. Western thinking, he believes, is a controlling, ruling kind of thinking. Plato in constructing his world of ideas becomes, according to Heidegger, the first technicistic philosopher. Cybernetics and information technology to Heidegger fulfill and at the same time negate Western philosophy (Heidegger, 1977; Schuurman, 1980).

In this light it is interesting to notice that Jacques Ellul interprets the history of philosophy differently, yet comes to the same conclusion. He says that philosophers have thought about the future as something positive and glorious, but that in the meantime technicistically inspired scientists, engineers, economists, and politicians, all with good intentions, have really gotten this wrong and prepared a negative and disappointing future. We have been betrayed by technology (Ellul, 1964, 1978, 1980), but this betrayal has been hidden through technological bluff (Ellul, 1990).

However that may be, some postmodernists such as Toulmin (1990) interpret the leading edge of multimedia, things like the digital city, as helping to legitimate a positive evaluation of our cybernetic age and our information society. For that reason the individualistic, not the universalistic, approach should get a chance. He thinks that the individualizing tendencies in the postmodern era are a sign of hope. He doesn't recognize, however, that our society as a postmodern society depends on the aberrations and problems of the technological culture, from which norms and values are disappearing. The same wishful thinking is found in the view of the theologian George Vahanian. The philosophers Rafaël Capurro and Heiner Hastedt also see positive connections between the information society and individuals with their personal interests. Similarly, the American philosopher Murray Bookchin (1980, 1982; Ferkiss, 1993) tries to synthesize postmodernism and naturalism.

Dewey, Staudinger, Horkheimer, Sachsse, Ihde, Tillich, Lyotard, Postman, and Rivers, as philosophers of culture, all emphasize the decisive stamp that technology has made on culture, and they all, in one way or another, identify technicism in one way or another as the main cause of our cultural problems.

36. Christian Faith and Culture

I come now to assessing the position of Christians in a technological culture. Generally speaking, Christians have accepted the ongoing technological development uncritically. The main reason for this is that they have been spellbound by the positive effects of technology, such as enriching material life, enhancing the duration of life, turning the tide in the struggle against poverty and illness, and so on. Are those things not all signs of the kingdom of God? Abraham Kuyper, for instance, stressed the need for technology. But he did not recognize the danger of technicism, the ideology of technology. Perhaps he was already a victim of it himself, when he wrote in his *Pro Rege* (1911) that the wonders of technology are greater than the miracles of Jesus. Although Simon Ridderbos has made the point in his study of Kuyper's theological view of culture that Kuyper's view of technology was complex, it cannot be denied that Kuyper warns more against dance, theater, and card playing, than he does against a technicistic trend in our culture.

The error of Kuyper, as a child of his time, was that he accepted technology as being free of problems. In a certain sense he accepted technology as neutral with respect to values. Kuyper, and most Christians with him, forgot the biblical warning that since the Fall, technology in the sense of the technology of craftsmen has been a power that leads astray, the more so as the ages roll by. Technology has more than once been a means to rival God, to make one's name on the earth and to build a culture without God—a Babel-culture (Schuurman, 1987). The origin is the Fall into sin: human autonomy connected to classical technicism. In its core, technicism wants to save human life without God. The Bible teaches us about Cain, about Lamech, about the building of the Tower of Babel, about Nebuchadnezzar, and so on. The Renaissance, Descartes, and the Enlightenment gave a new impetus to the modern form of technicism, namely absolutized scientific-technological control. The possibilities of modern technology have enhanced the ideology of technology. This ideology is mainly implicit. The gravity of this situation is often concealed by the numbing effects that emanate from a sweeping material prosperity.

Most of Kuyper's followers, too, have failed to evaluate technicism. Is it not astonishing that the Second Social Conference, held under Kuyper's leadership in the Netherlands in 1891, did not evaluate how technicism was guiding the development of technology? And can't the same be said of Christianity in general, when during the Conciliar Process in the churches the same theme is forgotten? Can it be that the biblical prophecy is not taken earnestly enough?

The last book of the Bible (Revelation chapters 17 and 18) speaks prophetically about the New Babel. That prophecy is a test for what I have said about the ideology of technology or technicism and its countless problems. When Christians accept technology only positively and uncritically—and, for instance, reduce the cultural mandate to a technological mandate—they are blind to what is really going on. They do not see that following the collapse of the collectivist materialism of Marxism experienced today in the countries of the former Soviet Union, the individualistic, liberal Western form of materialism will also collapse. Not seeing the religious background of our times, they are perhaps trying to christianize the secularized ideas of the Enlightenment. In reality they have closed their eyes to the central fact that human autonomy is concentrated in the scientific-technological control of everything. In any case, they seem unable to resist the unrestricted scientific-technological manipulation of reality.

People-without-God are trying to master God's creation. They want to realize a worldly paradise, a technological paradise. In the meantime the perspective of eternity is lost. Heaven is closed. Given the split between the divine world and earthly reality, they have set their hearts on technology and its results. This technicistic mind is the religious driving-force in our technological society and is enormously stimulated by welfare politics. If Christians are not aware of this religious background, they really lack the capacity to offer an adequate critique of our technological culture and to indicate the right, reforming normative direction to go. It is a pity that in the circles of Reformational Philosophy, we have spoken much about reforming philosophy and science but not enough so far about reforming culture. I think therefore that it will take a long time to analyze technological development critically and adequately. We are lagging far behind and as Christians have failed to develop a lifestyle suitable to a healthy culture.

In the technological culture the Christian worldview seems to have been reduced to a private belief lacking the capacity to reform culture. I am afraid that Christian thinking has indeed contributed little to reforming our technological culture. The reformation of scholarship and thought has not been placed in a broad cultural context, but has instead been reduced and restricted to the inner circle of the university. Although such an approach has its place, it ought to be broadened and deepened so as to include the whole of culture. That means fighting against technicism, scientism, and economism, and at the same time looking for the authentic meaning of all cultural activities and developing, in both practice and thought, a normative approach to each cultural field. This perspective or new direction will be possible with the help of the Holy Spirit under the guidance of the biblical groundmotive.

But before considering that, we need to see again more clearly what is going on in our culture. There has never been a time as technological as ours, but there has also never been an age so spiritually empty.

Materialism, consumerism, greediness, redundancy, and "killing time" are expressions of the withering of culture and spiritual death. Technicism is the deepest background of secularization today and of the loss of faith and the experience of the remoteness of God (Tillich, 1986). Our spiritual situation hinders us from seeing the depth and the impossibility of solving our countless cultural problems, including exhausting our raw materials, polluting the environment, fragmenting society in the form of individualism, increasing the gap between the rich and poor countries, structural unemployment, and so on.

Meanwhile, our obedience to the main norms of technological development such as that of technological perfection, the technological imperative (what can be made, ought to be made), and efficiency has brought our culture into a gigantic, dynamic process, which is at the same time destructive. If the Second and Third Worlds are brought up to the same dynamic level of material welfare as the West—and do they not have the right to it?—tomorrow's problems will grow enormously. The future seems not to end in a technological paradise but in a colossal chaos. The future of the earth appears to be at stake.

Andrew Feenberg (1991) wants to democratize technological development such that there will be a balance between efficiency, justice, and humanity. But when we look at the results of such a democratization we can better speak about *relegitimation* than about *reformation.* The same can be said about the ideas of Habermas, when he tries to disclose rationality in a social and democratic sense (1970). He wants to bring rationality under the guiding norms of information and communication.

It is understandable that new philosophic movements are opting for the other pole of our reformulated humanistic groundmotive, the adoration of nature. Christians ought also to be aware that these reactions do not offer a way out. The Norwegian philosopher Arne Naess (1991), with his concept of "deep ecology," is fully ecocentric. This movement desires in a certain sense to see a rebirth of animism: nature is divine in character and hence cannot be violated without punishment. A militant group within the movement of deep ecology is the "Earth First!" movement. They try to destroy the modern machinery used to control and gather what the earth produces.

Another reactionary group is guided by James Lovelock (1980) and Lynn Margulis (Ferkiss, 1993). The earth is seen and accepted as an organismic whole: Gaia. Gaia is involved or incorporated as a messiah. She can restore all that has been destroyed on earth, under the condition that humans accept themselves as a part of nature without a normative cultural calling to be stewards, without believing reality to be the creation of God the Creator, and without confessing the Fall into sin and Christ as Redeemer and Re-creator. Over against the technological paradise, these naturalistic thinkers are reclaiming the original paradise (McCormick, 1989).

37. The Biblical Groundmotive and Culture

The dialectic groundmotive of technological culture and nature stands in opposition to the biblical groundmotive of creation, fall, redemption through Christ in communion with the Holy Spirit and re-creation. This confrontation, which is founded neither in anthropocentrism nor ecocentrism (or naturalism), but in theocentrism or better Christocentrism, will dominate the next age. This conflict is concentrated in the struggle between the two groundmotives. The one is related to the *civitas terrena*, the terrestrial or technological paradise (dialectically related to "Mother Earth"), and the other one to the *civitas Dei*, the kingdom of God. Nevertheless, Babel (Rev. 17 and 18) parasitizes the New Jerusalem. The New Jerusalem is a gift of God, the fulfillment of creation as re-creation, in which human cultural activities are interwoven in a way that is not fully understandable (Rev. 21).

This gives a liberating and hopeful perspective. The dialectic in technological culture leaves room for the possibilities of a normative direction, even when the dialectic intensifies and assumes worldscale dimensions. Only when the whole of reality, including the relation between ourselves and nature or between us and technology, is seen as a unity founded in the Creation, disturbed in the Fall, restored in the Redemption, and soon to be fulfilled in the Re-creation as God's intervention, will a meaningful development of culture be possible.

The right, normative direction or meaningful perspective for cultural development stands over against both the accelerating technological culture with its problems and crisis, on the one side, and the naturalistic or ecocentric alternatives, on the other. Every field of cultural activity ought to be influenced by the biblical groundmotive. That groundmotive implies acknowledging that we are living between the Times, between the first and second coming of Christ. We are looking back in history and ought to be radically and integrally motivated by our calling for the future. Although the effects of the Fall will persist and be felt between the Times, we have to live in hope and expectation because the coming kingdom of God is the most decisive. Our position between the Times is fundamentally to struggle against every idol and to acknowledge God is the Creator, Redeemer, and Fulfiller. We are living in his creation, the meaning of which will be revealed, together with the divine mystery in history, in the kingdom of God.

Seen in that light, we must today primarily struggle against the power of technology and science, the power of absolutized scientific-technological control, and the influence of technological imperialism, in nearly every field in our culture. We ought to separate ourselves from the motives of secularized technological culture and turn to God in love, who loved us first in Christ, so that we can then return in love to our technological culture and our neighbors. Living in faith and hope and acting in love means that we are free both from technicism and its dialectic pole of naturalism.

Amidst centralizing, large-scale technological power devoted only to effectiveness and efficiency, we ought to differentiate and to welcome diversity. Effectiveness and efficiency are not the most important, but in the first place there ought to be love, righteousness, justice, service, mercy, thankfulness, and a readiness to make sacrifices (see also Schumacher, 1978).

It is an enduring consolation to know that on our own and by ourselves we cannot negate this meaning of creation, the kingdom of God. On the contrary, the fact that the kingdom of God is already on the way means that at any moment people may be converted and led once again to seek the kingdom, even in a technological society.

A radical and integral change of heart, a *metanoia* to God as followers of Christ—which is too often left entirely unmentioned in philosophic discussions—implies a different position in culture. That is the position of a "cross bearer" in the struggle against technicism and naturalism. That does not mean separation, alienation, or distance from technology. We could not live without it and we need it, but our hearts ought not be set on it. Technology should be no more than an individual and societal prosthesis; that is the authentic meaning of technology. Technology would then contribute meaningfully to all cultural sectors. In such a wholesome culture one could speak about the blessing of technology. Although not all problems would be solved, they would be bearable, and life in society would be liveable.

In conclusion, the philosophy of the cosmonomic idea, not as controlling but as faithful thinking, gives us the integral framework of norms, resting in God's law as an expression of his will (see §24). God's ordinances, which are revealed through the Scriptures and the creational order, provide the guide which we need (Monsma, 1986; Schuurman, 1993). That guide expresses the relativity of all cultural work, but at the same time the individual and institutional responsibility (Unger, 1982) to control culture and nature harmoniously. This struggle ought to be fought in the full armour of faith in the One Who holds all the power in heaven and on earth. When we fulfill Christ's power, the hidden ideology of technology will come to an end. Re-creation will overcome every distorting of creation. This final victory gives meaning to our cultural calling. We all should orient our cultural task to this perspective of eternity at the end of the twentieth century.

> In the world you have tribulation; but be of good cheer, I have overcome the world. (John 16:33)

Bibliography

Agassi, Joseph, *Technology: Philosophical and social aspects* (Hingham, MA: Reidel) 1986.

Altner, Günther, *Die Überlebenskrise in der Gegenwart* (Darmstadt: Wissenschaftliche Buchgesellschaft) 1988.

Beck, Heinrich, *Kultuurphilosophie der Technik: Perspektiven zu Technik, Menschheit, Zukunft* (Trier: Spee-Verlag) 1979.

Bohring, Gunther, *Technik im Kampf der Weltanschauungen* (Berlin: Deutscher Verlag d. Wissenschaften) 1976.

Bookchin, Murray, *The Ecology of Freedom: The emergence and dissolution of hierarchy* (Palo Alto, CA: Cheshire Books) 1982.

______, *Toward an Ecological Society* (Montreal: Black Rose Books) 1980.

Bos, Abraham P., *Wetenschap en zin-ervaring* (Amsterdam: VU Uitgeverij) 1985.

Brinkmann, Donald, *Mensch und Technik: Grundzüge einer Philosophie der Technik* (Bern: A. Franke) 1946.

Caplan, Alan L., "Ethical Engineers," 33/6 *Science, Technology and Human Values* (1980) 24-32.

Capurro, Rafaël, "Zur Computerethik—Ethische Fragen der Informationsgesellschaft", in H. Lenk and G. Ropohl, eds., *Technik und Ethik* (Stuttgart: Philip Reclam) 1987: 259-273.

Delschen, Karl-Heinz and Joachem Gieraths, *Philosophie der Technik: die Technikdiskussion in der Philosophie des 20 Jahrhunderts* (Frankfurt/Main: M. Diesterweg) 1982.

Dessauer, Friedrich, *Streit um die Technik* (Frankfurt/Main: Knecht) 1956.

Diels, Hermann, *Antike Technik*, 7th ed. (Leipzig: B.G. Teubner) 1924.

Dooyeweerd, Herman, "De gevaren van de geestelijke ontwapening der Christenheid op het gebied van de wetenschap," in J.H. de Goede, ed., *Geestelijk weerloos of weerbaar* (Amsterdam: Uitgevers-mij. Holland) 1936: 153-211.

______, "De transcendentale kritiek van het wijsgerig denken," 4 *Synthese* (1939) 314-339.

______, "De vier religieuze grondthema's in den ontwikkelingsgang van het wijsgerig denken van het Avondland," 6 *Philosophia Reformata* (1941) 161-179.

______, *Reformatie en Scholastiek in de Wijsbegeerte* (Franeker: Wever) 1949.

______, *A New Critique of Theoretical Thought*, trans. by David H. Freeman and William S. Young (Philadelphia: Presbyterian and Reformed) 1953.

______, *In the Twilight of Western Thought* (Philadelphia: Presbyterian and Reformed) 1960.

______, "The Secularization of Science," *International Reformed Bulletin* (1966, no. 9) 2-17.

______, *Roots of Western Culture: Pagan, secular and Christian options* (Toronto: Wedge) 1979.

Dijksterhuis, Edward Jan, *The Mechanization of the World Picture: Pythagoras to Newton*, trans. by C. Dikshoorn (Princeton, NJ: Princeton University Press) 1986.

Durbin, Paul T., ed., *Research in Philosophy and Technology* (Greenwich, CT: Jai Press) annually since 1978

______, ed., *Technology and Responsibility* (Hingham, MA: Reidel) 1987.

Ellul, Jacques, *The Technological Society* (New York: Knopf) 1964.

______, *Betrayal of the West* (New York: Seabury) 1978.

______, *The Technological System*, trans. by Joachim Neugroschel (New York: Continuum) 1980.

______, *The Technological Bluff* (Grand Rapids: Eerdmans) 1990.

Feenberg, Andrew, *Critical Theory of Technology* (New York: Oxford University Press) 1991.

Ferkiss, Victor, *Nature, Technology and Society: Cultural roots of the current environmental crisis* (London: Adamatine Press) 1993.

Ferré, Frederick, *Philosophy of Technology* (Englewood Cliffs, NJ: Prentice Hall) 1988.

______, *Hellfire and Lightning Rods* (New York: Maryknoll) 1993.

Forbes, R. J., *History of Science and Technology* (New York: Viking) 1963.

Goudzwaard, Bob, *Capitalism and Progress: A diagnosis of Western society*, trans. Josina Van Nuis Zylstra (Grand Rapids, MI: Eerdmans) 1980.

Grant, George Parkin, *Technology and Empire: Perspectives on North America* (Don Mills, Ontario: Stoddart Pub.) 1969.

Habermas, Jürgen, *Toward a Rational Society: Student protest, science, and politics*, trans. by Jeremy J. Shapiro (Boston: Beacon Press) 1970.

Hastedt, Heiner, *Aufklärung und Tecknik: Grundprobleme einer Ethik der Technik* (Frankfurt an Main: Suhrkamp) 1991.

Heidegger, Martin, *The Question Concerning Technology, and Other Essays* (New York: Harper and Row) 1977.

Hickman, Larry A., *John Dewey's Pragmatic Technology* (Bloomington, IN: Indiana University Press) 1990.

Hommes, Jakob, *Der Technische Eros: Das Wesen der materialistischen Geschichtsauffassung* (Freiburg: Herder) 1955.

Horkheimer, Max, *Critique of Instrumental Reason: Lectures and essays since the end of WWII*, trans. by Matthew J. O'Connell (Minneapolis, MN: Seabury Press) 1976.

Howe, Gunter, *Gott und die Technik: die Verantwortung der Christenheit für die wissenschaftlich-technische Welt* (Hamburg: Fürche-Verlag) 1971.

Huning, Alais, *Das Schaffen des Ingenieurs: Beitrage zu einer Philosophie der Technik* (Dusseldorf: VDI Verlag) 1987.

Ihde, Don, *Technics and Praxis*, Boston Studies in the Philosophy of Science, Vol. 24 (Kluwer: Norwell, MA) 1979.

______, *Existential Technics* (Albany: State University of New York Press) 1983.

______, "The Historical-ontological Priority of Technology over Science," in Larry Hickman, ed., *Philosophy, Technology and Human Affairs*

(1985) 196-211.

Jonas, Hans, *The Imperative of Responsibility: In search of an ethics for the technological age* (Chicago: University of Chicago Press) 1984.

______, *Technik, Medizen und Ethnik: zur Praxis des Prinzips Verantwortungen* (Frankfurt/Main: Insel Verlag) 1985.

Jungk, Robert, *The Nuclear State*, trans. by Eric Mosbancher (New York: Riverrun Press) 1984.

Kalsbeek, L., *Contours of a Christian Philosophy: An introduction to Herman Dooyeweerd's thought*, edited by Bernard and Josina Zylstra (Jordan Station, Ontario: Wedge) 1975.

Kapp, Ernest, *Grundlinien einer Philosophie der Technik* (Dusseldorf: Braunschewig) 1877.

Kranzberg, Melvin, ed., *Ethics in an Age of Pervasive Technology* (Colorado: Westview Press) 1980.

Kristensen, Brede, *Het verraad van de techniek: over de sociologie van Jacques Ellul* (Amsterdam: VU Uitgeverij) 1986.

Kuyper, Abraham, *Pro Rege*, I (Kampen: Kok) 1911.

Laszlo, Ervin, *The Systems View of the World: The natural philosophy of the new developments in the sciences* (New York: G. Braziller) 1972.

______, *Introduction to Systems Philosophy: Toward a new paradigm of contemporary thought* (New York: Harper and Row) 1972.

______, *A Strategy for the Future: The system approach to world order* (New York: G. Braziller) 1974.

Lenk, Hans, *Technokratie als Ideologie: Sozialphilosophische Beitrage zu einen politischen Dilemma* (Stuttgart: W. Kohlhammer) 1973.

______, *Zur Sozialphilosophie der Technik* (Frankfurt/Main: Suhrkamp) 1982.

Lenk, Hans, ed., *Techne, Technik, Technologie: Philos. Perspektiven* (Pullach: Verlag Dokumentation) 1973.

Lovelock, James E., *Gaia: A new look at life on Earth* (New York: Oxford University Press) 1980

Lyotard, Jean-F., *The Inhuman: Reflections on time* (Stanford, CA: Stanford University Press) 1992.

Maes, A.J.M.M., ed., *Vooruitkijken naar vooruitgaan: Technologie in de toekomst* (Den Haag: Directie Algemeen Technologiebeleid) 1993.

Marx, Leo, "The Idea of 'Technology' and Postmodern Pessimism," in Y. Ezrahi *et al.* eds., *Technology, Pessimism, and Postmodernism* (Dordrecht: Kluwer Academic Publishers) 1994: 11-29.

McCormick, John, *Reclaiming Paradise: The global environmental movement* (Bloomington: Indiana University Press) 1989.

Meyer, Herman J., *Die Technisierung der Welt: Herkunft, Wesen und Gefahren* (Tubingen: Niemeyer) 1961.

Minsky, Marvin, *The Society of Mind* (New York: Simon and Schuster) 1988.

Mitcham, Carl and Jim Grote, eds., *Theology and Technology: Essays in Christian analysis and exegesis* (Lanham, Md.: University Press of America) 1984.

Monsma, Stephen V., ed., *Responsible Technology: A Christian perspective* (Grand Rapids: Eerdmans) 1986.

Müller-Schwefe, Hans Rudolf, *Technik und Glaube: eine permanente Herausforderung* (Mainz: Matthias-Grunewald-Verlag) 1971.

Naess, Arne, *Ecology, Community and Lifestyle: Outline of an ecosophy* (Cambridge: Cambridge University Press) 1991.

Pippin, R.B., "On the Notion of Technology as Ideology," in Y. Ezrahi *et al.* eds., *Technology, Pessimism, and Postmodernism* (Dordrecht: Kluwer Academic Publishers) 1994: 93-114.

Postman, Neil, *Technopoloy* (New York: Knopf) 1992.

Raby, W.H., *Über Sinn und Grenzen der Technik* (Koln) 1966.

Rapp, Friedrich, ed., *Contributions to a Philosophy of Technology: studies in the structure of thinking in the technological sciences* (Hingham, MA: Reidel) 1974.

Rapp, Friedrich, *Analytical Philosophy of Technology*, trans. by Stanley R. Carpenter and Theodor Langenbruch (Hingham, MA: Reidel) 1981.

Rapp, Friedrich and Paul T. Durbin, eds., *Philosophy and Technology* (Hingham, MA: Reidel) 1983.

Ridderbos, Simon J., *De theologische cultuurbeschouwing van Abraham Kuyper* (Kampen: Kok) 1947.

Rivers, Theodore J., *Contra Technologiam—The crisis of value in a technological age* (New York: University Press of America) 1993.

Rogers, Gordon F. C., *The Nature of Engineering: A philosophy of technology* (London: Macmillan Press) 1983.

Sachsse, Hans, *Technik und Verantwortung: Probleme der Ethik im Technischen Zeitalter* (Freiburg: Rombach) 1972.

______, *Anthropologie der Technik: Ein Beitrag zur Stellung des Menschen in der Welt* (Braunschweig: Viewg) 1978.

Schumacher, Ernest F., *Small is Beautiful: Economics as if people mattered* (New York: Harper and Row) 1973.

______, *A Guide for the Perplexed* (New York: Harper and Row) 1978.

Schuurman, Egbert, *Techniek: middel of Moloch?* 2nd ed. (Kampen: Kok) 1980.

______, *Technology and the Future: A philosophical challenge*, trans. by Herbert D. Morton (Toronto: Wedge Publishing Foundation) 1980.

______, *Reflections on the Technological Society* 2nd ed. (Toronto: Wedge Pub. Foundation) 1982.

______, "A Christian Philosophical Perspective on Technology," in Carl Mitcham and Jim Grote, eds., *Theology and Technology* (New York: University Press of America) 1984: 107-123.

______, *Tussen technische overmacht en menselijke onmacht: Verantwoordelijkheid in een technische maatschappij* (Kampen: Kok) 1985.

______, *Christians in Babel* (Jordan Station, Ontario: Paideia Press) 1987.

______, "The Modern Babylonian Culture," in P. Durbin, ed., *Technology and Responsibility* (Dordrecht: Reidel) 1987: 229-243.

______, "The Agricultural Crisis in Context: A reformational philosophical perspective," XVII *Pro Rege* (September, 1989) 2-14.

______, *Het "Technische Paradijs": Om de gebrokenheid van heel de schepping* (Kampen: Kok/Voorhoeve) 1989.

______, *The Future: Our choice or God's gift?* (New Zealand: Exile Publications) 1991.

______, "Technicism and the Dynamics of Creation," 58 *Philosophia Reformata* (1993) 185-191.

Sizoo, A., *Techniek in de oudheid* (Kampen: Kok) 1961.

Staudinger, Hugo and Wolfgang Behler, eds., *Chance und Risiko der Gegen-*

wart: Eine kritische Analyse der wissenschaftlichtechnischen Welt (Paderborn: Schoningh) 1976.

Stork, H., *Einführung in die Philosophie der Technik* (Darmstadt: Wissenschaftliche Buchgesellschaft) 1977.

Strijbos, Sytze, *Het technische wereldbeeld: Een wijsgerig onderzoek van het systeem denken* (Amsterdam: Buijten & Schipperheijn) 1988.

Tillich, Paul, *The Spiritual Situation in Our Technical Society* (New York: Scribner) 1986.

Toulmin, Stephen, *Cosmopolis—The hidden agenda of modernity* (New York: Free Press) 1990.

Unger, Stephen H., *Controlling Technology: Ethics and the responsible engineer* (New York: Holt, Rinehart, and Winston) 1982.

Vahanian, George, "Christliche Religion und Kultur," in A. Hertz *et al.* eds., *Handbuch der christlichen Ethik* (Freiburg: Herder) 1984: 439-453.

Van Griethuysen, A. J., ed., *Grenzen aan techniek* (Alphen aan den Rijn, The Netherlands: Samsom) 1989.

Van Melsen, Andrew G. M., *Science and Technology* (Pittsburg: Duquesne University Press) 1961.

______, *Science and Responsibility*, trans. by Henry J. Koren (Pittsburg: Duquesne University Press) 1970.

Van Peursen, Cornelius A., *et al.* eds., *Informatie: een interdisciplinaire studie* (Utrecht: Spectrum) 1968.

______, *Cultuur in Stroomversnelling* (Amsterdam: Elsevier) 1985.

______, *Filosofie van de wetenschappen* (Leiden: Nijhoff) 1987.

Van Riessen, Hendrik, *Filosofie en Techniek* (Kampen: Kok) 1949.

______, *The Society of the Future*, trans by David H. Freeman (Philadelphia: Presbyterian and Reformed) 1957.

______, *Wijsbegeerte* (Kampen: Kok) 1970.

Van Woudenberg, René, *Gelovend denken: Inleiding tot een christelijke filosofie* (Amsterdam: Buijten en Schipperheijn) 1992.

Vermeersch, Emile, *De ogen van de panda—Een milieu-filosofisch essay* (Brugge: Van der Wiele) 1980.

Von Schilling, Kurt, *Philosophie der Technik: die geistige Entwicklung der*

Menschheit von den Anfängen bis zur Gegenwart (Herford: Maximilian-Verlag Kurt Schober) 1968.

Von Weizsäcker, Carl F., *The Relevance of Science* (London: Collins) 1964.

Weizenbaum, Joseph, *Computer Power and Human Reason: From judgment to calculation* (San Francisco: W.H. Freeman) 1976.

Winner, Langdom, *Autonomous Technology: Technics-out-of-control as a theme in political thought* (Cambridge, MA: MIT Press) 1977.

Wuthnow, Robert, *The Restructuring of American Religion: Society and faith since World War II* (Princeton: Princeton University Press) 1988.

Index of Persons and Subject-Matter

About the Author

Egbert Schuurman (1937) studied civil engineering at the Technological University of Delft and philosophy at the Free University of Amsterdam and is presently Professor in Reformational Philosophy at the Technological Universities of Delft and Eindhoven and at the Agricultural University of Wageningen. He is also a member of the Senate of the Dutch Parliament, chairman of the Prof. Dr. G.A. Lindeboom Institute's Center for Medical Ethics, and doctor *honoris causa* of the Potchefstroom Christian University in South Africa. His main publications in English are: *Technology and the Future: A philosophical challenge* (Toronto: Wedge, 1980), *Reflections on the Technological Society* (Toronto: Wedge, 1983), *Technology in Christian-Philosophical Perspective* (Potchefstroom, 1981), *Information Society: Impoverishment or enrichment of culture?* Potchefstroom, 1984), *Christians in Babel* (Ontario: Paideia Press, 1987), *The Future: Our choice or God's gift?* (Wellington, New Zealand: Exile Publications, 1991), and co-author of *Responsible Technology: A Christian perspective* (Grand Rapids: Eerdmans, 1986). Several of his books have been translated into Korean, Japanese and Chinese.